The Health Hazards of NOT Going Nuclear

The Health Hazards of NOT Going Nuclear

By

Petr Beckmann

Electrical Engineering Department
University of Colorado

THE GOLEM PRESS

Library of Congress in Cataloging in Publication Data:
Beckmann, Petr.
The health hazards of NOT going nuclear.
Includes bibliographical references and index.
1. Atomic power plants – Hygenic aspects.
2. Atomic power plants – Environmental aspects.
3. Electric power plants – Hygenic aspects.
4. Electric power plants – Environmental aspects.
1. Title
RA569.B43 363 76-12720
ISBN o-911762-17-5

First printing July 1976
Second printing September 1976
Third printing January 1977
Fourth printing August 1977
Fifth printing November 1977
Sixth printing May 1978
Seventh printing March 1979
Eighth printing October 1979
Ninth printing June 1980
Tenth printing March 1985

Printed in the U.S.A.

Library of Congress Catalog Card Number: 76-12720

$7.95 postpaid from

The Golem Press
Box 1342, Boulder, Colorado 80306

To
Ralph Nader
and all who worship
the water he walks on

A Post-Three Mile Island Preface

When this book was first published 3½ years ago, it didn't have a preface. But since then we have had the Three Mile Island Grand Disaster — history's only major disaster with a toll of zero dead, zero injured, and zero diseased.

How much of this book had to be rewritten as a result of this Grand Disaster?

Not a solitary line; not a solitary word; not a solitary i-dot.

On the contrary, the reader is cordially invited to use the Grand Disaster as an experimental test of what this book asserts, in particular, as a test of the central question of whether the zero casualty toll was "a lucky near miss" or whether it was a logical consequence of the two fundamental pillars of nuclear safety: the defense in depth and the slow progress of a nuclear accident.

The defense in depth, and why no other energy facility of equal size can have one, is described in the book; but perhaps the slow time scale of a nuclear accident should have been more strongly emphasized, so here is an illustration from the TMI episode: Within hours from the beginning of the accident [Hours? How long does it take for an oil tanker to blow up?], the industry had flown in teams of experts; one such team engaged in almost Naderite "what-if" fantasies. What if the pump now slowly cooling the core fails? We use the other primary loop. What if that fails, too? We still have the ECCS. What if both loops fail because the power fails? We have a diesel stand-by generator. What if that fails, too? Let's fly in another, just in case. And they did. (It was never needed.)

What 843 MW facility, other than nuclear, gives you that kind of time to take countermeasures? What other 843 MW facility will contain a chain of five independent, horrible failures — human and mechanical — without the loss of a single life? How do you evacuate the population when a dam breaks? How do you take preventive measures while a gasoline refinery blows up?

Yet Three Mile Island did not become what it should have — a gigantic field test of nuclear safety. It became the bugle call for the biggest brainwashing campaign in American history.

"Scientists told us an accident of this type was virtually impossible, but now it has happened," is an assertion proved false by this book, published almost 3 years before the TMI accident. An accident like TMI, which resulted in nothing but property damage (and would not have resulted even in that if an operator had not switched off the correctly functioning automatic safeguard) always had, and continues to have, a significant probability; what is highly improbable is an accident with large-scale loss of life, such as happens very frequently with other energy sources. Since this book was published, dam disasters have killed thousands of people (at least 2,000 in India in August 1979); many hundreds have died in explosions and fires of gas, oil, butane, gasoline, and other fuels; and tens of thousands have died premature deaths, in the US alone, due to the use of coal. Many of these deaths could have been avoided by using nuclear power as a preferred source of electricity; but who cares about safety or public health? Not the politicians, nor the self-anointed Messiahs. The real dangers continue to be covered up; it is, after all, the *perceived* dangers from which political capital can be made. So let the suckers die for the greater glory of Kennedy, Udall, Hart, Ottinger, Jerry Brown and Tom Fonda.

A meltdown? It was never even close at TMI; and this book will tell you why it would not have been the end of the world even if there had been one.

Radiation? The average dose received by people in the neighborhood of TMI due to the accident was one millirem; the maximum anybody could have received was about 80 mrem. (Just by staying in Colorado since this book was published, I have received 350 mrem more than if I had moved to Harrisburg, Pa.) Of the population now living within 50 miles of the TMI site, approximately 350,000 would eventually have died of cancer if there had been no accident. That number will now be increased by one, possibly by two.

And those are the only deaths to be expected, say the official figures.

Wrong! Many more than that will die due to the accident, and some of them have already died. What is being covered up is that TMI Unit 2, now crippled, was *saving* lives which are now being lost in producing electricity by far less safe power sources. Most of the replacement power is coal-fired, and the detailed studies by Brookhaven National Laboratory published in 1978-9 have found more accurate estimates of the death toll due to coal-caused air pollution:* The median is 74 per 1000 MW coal-fired electric power (this book said between 40 and 100). That makes

* *The Direct Use of Coal*, Office of Technology Assessment, April 1979; $7 from Government Printing Office, Washington, DC 20402; stock no. 052-003-00664-2.

about 62 dead for the missing 843 MW of TMI Unit 2 by air pollution alone, or more than 1 death per week.

That cannot be helped, of course; Unit 2 had an accident. But what about TMI Unit 1? This *could* be saving more than 1 life a week, but it is shut down for no good reason other than to get more political mileage out of the accident. For since this book was published, the NRC has changed, at least in part, from a commission of timid experts to a commission of political demagogues: I refer to Commissioner Gilinsky, and above all, to Carter appointee Peter Bradford, a lawyer formerly working for Ralph Nader's corporation-baiting conglomerate. It is people like Bradford who keep nuclear plants shut down for every conceivable excuse. In August 1979, Bradford acknowledged that he was aware of coal being riskier than nuclear power; so he must know about the 1 death a week due to TMI Unit 1 standing idle. But politics is politics; what's a few more widows?

Not all casualties of TMI were fatal, or even regrettable. One such casualty was the theory of the inevitable failure of the Emergency Core Cooling System as diligently expounded by Prof. Henry Kendall and his Union of Concerned "Scientists." He had built a lucrative career on the alleged malfunction of the ECCS, which had never been tested except in what he considered artificially staged acts of whitewash. At TMI all of his carefully nursed predictions of doom were shattered in a fraction of a second as the ECCS came in immediately, reliably and perfectly. (It was later switched off by human error.) Deeply wounded by this insult, he called for evacuation (with its inevitable deaths by heart attacks of the elderly, traffic accidents, etc.) at a time when there was no danger of a massive radioactive release. He also predicted that a meltdown will cause 100,000 deaths, will slaughter children under 12, will massively pollute rivers, and lead to all kinds of horrors that would scare the pants off Count Dracula; at the same time his organization assured us that they did not want to halt nuclear power completely.

Now if I believed, as Kendall purports to believe, that nuclear power could result in widely contaminated landscapes littered with corpses, I would certainly turn against it; but then, perhaps I am not quite as tough as Kendall. What kind of technology *would* turn him off? Artificial earthquakes with rockets releasing the bubonic plague?

Some other events in the last three years deserve comment. The methodology of the Rasmussen study — though you would never believe it from the reports by the brainwashers in the national news media — has been upheld and endorsed by the NRC (lawyer Bradford was in the minority on that one); it merely disassociated itself from the Executive

Summary of the report and could not endorse the exact values of the probabilities as calculated in it — as Prof. Lewis, chairman of the review group, testified before a Congressional Committee, the real probabilities could well be more *favorable* to nuclear power.

The attacks on low-level radiation have reached the borders of sanity — and here I do not merely allude to Dr Ernest Sternglass, whose "findings" on mental retardation, infant mortality and decline in test scores have grown so wild that they have become an embarrassment to the antinuclear lobby. As in so many other cases, I must refer the reader to my monthly newsletter *Access to Energy** to keep abreast of the latest facts and fictions; let it only be mentioned that Dr Mancuso published his report *after* his contract was left to expire without renewal because he had not produced results of any kind in many years, and that Dr Najarian was dismissed by a fuming Sen. Kennedy in June 1979 after his testimony on alleged increased cancer incidence among nuclear shipyard workers was demolished by medical experts.

For the rest, the antinuclear thrust has not changed direction; it has only intensified and the antinuclear movement has become part of the political establishment. My contention that this movement is largely fed by members of a class who want to freeze society in the state where they occupy privileged positions has received much supportive evidence, though I still do not claim it to be the only explanation.

One of the important, though usually unreported, developments of the last three years is the extent to which the victims of such a policy — those who are denied upward social mobility by the no-growth advocates — have begun to struggle against it. The starting point may well have been the NAACP's statement on energy in December 1977; by now there is a growing grassroot pro-energy movement which seems to be unnoticed, misread or ignored by the current set of politicians. The next set may well reach the troughs of power by recognizing what their predecessors overlooked, and though it is improbable that they will be more moral or honest, the prospect of the current mismanagers being turned out and sent packing is still something to be looked forward to.

Among the more telling glimpses of the antinuclear movement's concern with social engineering rather than safety is a statement by Amory

* $12 for 12 monthly issues from Box 2298, Boulder, Colorado 80306. (Price will not be increased if the federal budget is balanced.)

Special subjects are treated by this author in his *Different Drummer* booklets such as 1. *Nuclear proliferation — how to blunder into it*; 3. *Small is Beautiful? Economics as if only SOME people mattered*; 6. *Why "soft" technology will not be America's energy salvation*; 7. *The non-problem of nuclear wastes*. $2 each from Golem Press, Box 1342, Boulder, CO 80306.

B. Lovins, a college dropout variously described by his mentors as physicist, economist and genius:

"If you ask me, it would be little short of disastrous for us to discover a source of clean, cheap, abundant energy because of what we would do with it."

On the other hand, few people have put the issue as succinctly as Prof. J.H. Fremlin of the University of Birmingham, England, in his commentary (*Biology*, vol. 25, no. 4, 1979) on the "Windscale Inquiry" (which considered the issue of reprocessing nuclear fuel at Windscale in northern England — the project has since been approved and is now operating):

"I do not regard it as my job to tell people what they ought to do. But I can summarize the whole discussion on safety at Windscale and elsewhere by saying that if you think the most important thing is that as few people as possible should be killed, then you will press for a switch as rapidly as possible from fossil fuels to nuclear power. If on the other hand you think it most important that as few as possible people should be frightened, then you will press for the abandonment of all nuclear power in favour of fossil fuels."

I hope this book will be useful for both these schools of minimization.

P.B.

Boulder, Colorado
Fall 1979

Contents

1

The Nuclear Monologue

How many atomic explosions in our cities would you accept before deciding that nuclear power is not safe — no complexities, just a number!

Question posed to AEC Commissioner Doub at Ralph Nader's "Critical Mass" meeting, November 1974.

What is remarkable about the above quotation is not so much its loutish arrogance, nor even the speaker's preference of political sloganeering over technical complexities; what is remarkable is the speaker's abysmal ignorance.

For the speaker (or heckler) has based his "question" on two totally false premises: First, that anybody of any consequence ever claimed that nuclear power is safe, and second, that an atomic explosion in a nuclear plant is possible.

Both implications are utterly false. There is no such thing as safe energy conversion on a large scale; it is almost a contradiction in terms. Energy is the capacity for doing work, and as long as man is fallible, there is always the possibility that it will do the wrong kind of work; to ask for absolutely safe energy, therefore, is much the same thing as asking for incombustible fuel.

This book never tries to make the point that nuclear power is safe; the point it makes is that it is far safer than any other form of large-scale energy conversion yet invented.

The other implication, that of an atomic explosion in a nuclear plant is even more preposterous, for such an explosion is physically impossible. Not highly improbable, but utterly impossible: An explosive nuclear chain reaction is no more feasible in the type of uranium used as power plant fuel than it is in chewing gum or pickled cucumbers.

And yet these two fallacies, ludicrous as they may be, are not the only ones that have become widespread among the public. Equally ludicrous are the falsehoods that nuclear power is less reliable than fossil-fired power, that insurance companies are unwilling to insure nuclear plants for liability, that nuclear power will lead to a "radioactive society," and many more such superstitions that will be examined here.

Nor are these superstitions shared and disseminated merely by a self-destructive intellectual elite gone berserk in its hatred of the system that elevated it to its present position. These myths have made inroads among honest citizens concerned about the safety of their communities. Even some scientists (though almost none in the field of nuclear power) have become scared of nuclear power.

Politicians, who in the late sixties changed their vocabulary from justice and motherhood to ecology and environment, are perking up their ears: Ever ready to cater to prejudices that will bring votes, they are probing whether *nuclear* can be made into a dirty word, as dirty as *profits* maybe, so that they can gallantly wage an anti-nuclear campaign to save the widows and orphans from the greedy corporations.

Several states have already passed legislation curbing the growth of nuclear power; in June 1976, California will vote on the "Nuclear Initiative," a piece of demagoguery disguised as an initiative for better safeguards, but in fact posing irrational conditions that would effectively prohibit nuclear power in California. Win or lose, the superstition mongers will go on to other states to crusade against nuclear power; to crusade, consciously or not, for more deaths by mining accidents, Black Lung, air pollution and chemical explosions; to crusade, consciously or not, for increased American dependence on medieval sheikdoms and other unstable dictatorships.

The so-called nuclear debate is replete with myths, distortions and outright falsehoods; but it is compounded by the most exasperating of them all, the myth that there is a nuclear debate at all.

What debate?

There is no debate, only a monologue. There have been almost no reasoned debates between proponents and opponents of nuclear power; what there has been in abundance is coverage, especially by the TV networks, of puerile "what-if" fantasies limited exclusively to nuclear power, never applied to fossil-burning plants or other energy sources. There has been, and continues to be, excessive coverage of the hit-and-run tactics of Ralph Nader, whose ignorance of nuclear power is matched only by his arrogance in discussing it. There are the insiduous "documentaries" by the TV networks that pretend to give a balanced view by balancing the truth against a lie; or worse, by persistently manipulating the viewer with the-truth-but-not-the-whole-truth insinuations ("last week, Walter Cronkite was sober for three straight days" would be a counter-example). Above all, the media keep playing on the psychological association of "nuclear" and "bomb," which makes as much sense as the association of "electric" and "chair."

The national press is little better; with few exceptions, it has joined in the fashionable sport of slandering nuclear power and fanning the hysteria. And this does not just refer to the *Mother Earth News* or the Naderite press; it includes such publications as *The Wall Street Journal* and *Business Week*, which can hardly be considered periodicals of the counter-culture. And the wire services — Associated and United Press — feed the local papers with centrally generated distortions.

On the other side there are, or rather should be, the scientific community, the nuclear industry, and the utilities. But they have been silent partners to the monologue; in part, because they do not speak up, in part, because they cannot make themselves heard.

Among those that have not spoken up, at least not effectively, is the nuclear industry. Ralph Nader has become the laughing stock of that industry, but until recently, they have been laughing all by themselves. Westinghouse has published a series of advertisements trying to explain the facts, but the credibility of advertising is low. The Westinghouse series was very well done and entirely truthful — but who believes Westinghouse any more than Anacin? And Anacin, after all, does not expect to be believed or disbelieved; it just batters the brain with meaningless lyrics. (What, pray, is the meaning of "In the final analysis only Anacin hits and holds the highest level"?)

The Atomic Energy Commission let itself be crowded into the defensive and in its final years hoped to appease its critics by increasing the rigor of nuclear safeguards, in some cases to the borderline of the absurd. The Nuclear Regulatory Commission, which took over the regulatory program of the AEC (the research program passed to ERDA), has often continued in this policy of appeasement. Typical of this policy is its decision to delay plutonium recycling, a decision which is not only detrimental to the economics of nuclear power, but also to its safety, for nuclear wastes, including plutonium compounds, are now piling up at individual plant sites while the NRC polishes its image as a watchdog of safety.

The utilities are, naturally, aware of the excellent economics, as well as the safety, of nuclear power; at the time of writing, the price of nuclear fuel is about 6 times lower than that of the average fossil fuel needed to generate the same amount of electrical energy (even though the prices of oil and gas are controlled at unrealistically low levels), and even when the depreciation costs of the more expensive nuclear plants are considered, the overall cost of nuclear power amounts to between 50% and 80% of the cost of fossil-generated power.

But utilities throughout the country are in a near disastrous bind of capital shortage and they have canceled or deferred about half the orders for new and badly needed nuclear capacity; and as behooves an industry that is shackled hand and foot by government regulations, price controls, rate controls, and politicized Public Utility Commissions, it keeps its mouth shut. Only very recently, pressed against the wall by the nuclear initiatives, have some utilities begun to speak up.

Moreover, the utilities and their PR departments are afraid of "knocking coal." Most of their capacity, and often all of it, is fossil-fired. By pointing out that nuclear power is safer than fossil-fired power, they would admit that their present power generation is not the safest possible; and well knowing that if nuclear power is banned, the activists will next turn against coal (for they are against *all* large-scale energy conversion, hoping to force their recommended lifestyle on everybody else), the utilities fear that if they lose in the nuclear initiatives, their own arguments will be turned against them, and they prefer to be quiet on the point.

This is, in my opinion, a mistake. It amounts to giving up much of the most effective, indeed, the *only* weapon the nuclear supporters have: the truth. Even if there were a genuine debate and the battle were even, the truth should be the one and only weapon; but to forego it in a situation without equal access to the mass media, to forego it in a situation which enables the activists to make 30-second statements that it takes half-hour lectures to refute, is suicidal.

Besides which, there is no need to "knock coal." Though fossil fuels are far more dangerous than nuclear power, they save far more lives than they take, as does any form of large-scale energy conversion — one need merely compare the public-health statistics of an advanced, energy-intensive economy with those of a backward economy, no matter whether in the US in the past or elsewhere in the present. And the need for energy to maintain the US standard of public health (as well as the general standard of living) is such that we cannot afford, or even quickly achieve, the exclusive use of the safest; we must settle for the second-safest and third-safest as well. When the entire picture is considered, fossils and hydropower are safe to a very high degree (certainly much safer than the small-scale gadgets advocated by the primitivists), and the reader of this book is earnestly requested *never to forget that the point of this book is not to argue how dangerous fossils and hydropower are, but to show that nuclear power is safer.*

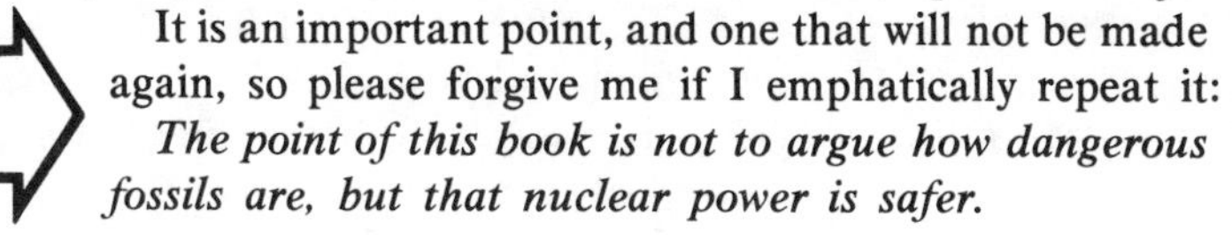

It is an important point, and one that will not be made again, so please forgive me if I emphatically repeat it: *The point of this book is not to argue how dangerous fossils are, but that nuclear power is safer.*

And finally, there is the scientific community. "Scientists," we are invariably told, "are sharply divided on the issue of nuclear power." What makes this statement so exasperating is that, in a certain sense, it is true. What the statement does not say is how the dividing line runs: The opponents of nuclear power are recruited from the ranks of entomologists, anthropologists, biologists, neurologists, chemists, and other non-nuclear disciplines; but there are very few nuclear engineers of any repute among them. Typically, the three nuclear engineers who resigned from General Electric in San Jose, Calif., in February 1976, were members of a para-religious organization[1] (the Creative Initiative Foundation, which teaches that "God did not make plutonium, and therefore it is evil," inviting the question whether Our Daily Bread is evil, too). Though the well-heeled C.I.F. apparently promised financial security to others who would join in this propaganda gimmick, they were unable to recruit more than 3 of the 480 other engineers at the same level working for GE's Nuclear Division,[2] which is not surprising, since most nuclear engineers are aware of the fact that the alternatives of nuclear power endanger, and indeed take, more lives.

Among nuclear physicists, there is only a handful of critics of nuclear power. Best known among them is Henry W. Kendall of the Massachusetts Institute of Technology, who is active in the Union of Concerned Scientists. Kendall, who still nurses an old grudge against the AEC for rebuffing him, now acts as an adviser to Ralph Nader; but not even Kendall can be persuaded by Nader to endorse a nuclear moratorium, nor was Kendall able, when challenged by this writer, to name a method of generating power safer than nuclear.

The typical "distinguished scientist" opposing nuclear power, sometimes a Nobel Prize winner, is a man of quite a different ilk. He has two

---

[1] Superior figures refer to notes at the end of the book.

salient characteristics: First, he made his name in a field unconnected with nuclear power, and second, he has a penchant for embracing political causes.

Linus Pauling, for example, won his Nobel Prize in Chemistry on a subject that has nothing to do with nuclear power, and he is known to the general public mainly for his escapades such as posing as a one-man picket in front of the White House to protest the Viet Nam war. His proposed cure of the common cold (also outside his original field of expertise) has recently been disproved, and there is no reason to believe that he knows more about colds than coolants.

Hannes Alfven obtained his Nobel Prize for his contribution to plasma physics, particularly as applied to the ionosphere, a set of layers in the atmosphere from 50 to 500 miles above the surface of the earth. His statements on the dangers of nuclear power show that he has not only little understanding of nuclear power, but that he has no understanding of the concept of safety: He evidently believes in the existence of absolute safety and requires it for nuclear, but not for any other kind of power.

Barry Commoner is a biologist who has done important work on genetic mutations induced by carcinogens in bacteria, but he is better known to the public as a doomsday prophet, an opponent of economic growth, an advocate of nationalizing the railroads and the energy industries, and a crusader against "big business" who has recently endorsed Marxist economics.[3]

There are many other scientists in this group — scientists who have distinguished themselves in a field far removed from nuclear power, and who have embraced political causes for which to crusade.

Not included in this group are men like Ehrlich, Tamplin or Goffman, who can at best be called ex-scientists. Mediocrities in their own fields, they seem to have tried for a quicker way to glory, and they now specialize in horror stories that are reprinted in Sunday supplements to scare the gullible. The science fiction produced by Tamplin, Goffman, Sternglass and others has been refuted many times by scientific committees and is too ridiculous for all but the politicized environmental organizations. As for Dr Ehrlich, it is difficult to talk about his "ignorance" of physics, for if he knew nothing about it, it would be a marked improvement; his misunderstanding of thermodynamics, for example, is shocking.[4]

Returning from these ex-scientists to the scientists, one may say that the opponents of nuclear power are not only drawn from the ranks of other disciplines, but they represent a minute fraction of the scientific community. Typically, this small minority sent a petition to slow down nuclear power to President Ford in August 1975 on the 30th anniversary of the Hiroshima nuclear bomb explosion. That in itself shows that the petition was an unprofessional political gimmick, since nuclear bombs are no more related to nuclear power than electric power is related to the electric chair.

Out of a total of 200,000 scientists (counting only the physical and life sciences, and only at the universities), the Union of Concerned Scientists mailed its petition to 15,000 names: members of the American Federation of Scientists and subscribers to the *Bulletin of Atomic Scientists*, both organizations which have long since forsaken science for politics, and the latter blatantly anti-nuclear. It was much like asking the National Rifle Association what it thought of gun control; but even so, the perpetrators of this gimmick were able to gather only 2,300 signatures, or 0.3% of the 770,000 scientists in the physical and life sciences,[13] and that does not include workers in the social sciences, who are usually prominent in this type of petition.

Kendall's petition, then, was a flop in the scientific community, and a total fiasco as far as the hard sciences are concerned. Yet the *Washington Post*, for example, reported on this petition under the headline *Scientists Urge Slowdown of Nuclear Power.* And hundreds, perhaps thousands, of local papers that reprint this type of story from the *Post* or *New York Times* have used the same headline above the same pap.

As late as March 1976, the *Christian Science Monitor*, while admitting that the three engineers who had resigned from General Electric were members of a para-religious anti-nuclear group, claimed that scientists were "split down the middle" on the issue of nuclear power and pointed to the 2,300 scientists who had grave doubts about it.

But what about the other side? The scientists defending nuclear power are different from their opponents in almost every respect. They have no political ambitions; they are as yet unorganized; they do not ta!k about vague dangers, but about hard numbers; they receive virtually no media exposure; they are far less vociferous; but above all, most of them are men who know nuclear power and nuclear hazards, not from political meetings, but from direct and immediate experience.

There are as yet few scientists who have taken to the pen to address the broader public outside the readership of the professional literature to defend nuclear power. But unlike their neurologist and entymologist opponents they do know what they are talking about. Dr Ralph Lapp, one of the most successful popularizers of nuclear science, is a high-energy physics researcher who has been in the nuclear field for more than 30 years, starting out as a researcher under Nobel Prize winner Arthur H. Compton and successively holding a long series of distinguished appointments at various universities and government agencies, including the post of Assistant Director of the Argonne National Laboratory. He is now an energy and nuclear consultant, and senior member of a non-governmental nuclear corporation.

Dr R. Philip Hammond, a nuclear physicist with more than 30 years experience with reactors and fission wastes, is a former professor at the University of California at Los Angeles and now a consultant in the energy field. Where Alfven and Kendall (let alone Tamplin or Ehrlich) let their imaginations run wild with sick "what if" scenarios of nuclear accidents, Dr. Hammond talks like this:

"If I had to contend with such material [radioactive material after a major meltdown accident] — and I have had some first-hand experience in cleaning up radioactive spills — I cannot think of a place where I would prefer to have it than underground... I would be glad to tackle the job of drilling into the spilled fuel and bring it up in small bits for recovery. This could be done safely and completely."

At the time of writing, there have been only a handful of petitions or formal declarations in support of nuclear power (for a good thing does not need such proclamations until a malicious campaign has started against it); but when such proclamations have been made in answer to anti-nuclear attacks, their authors had no trouble finding Physics Nobel Prize winners and scientists directly working in the nuclear field. The 33 outstanding scientists who signed a statement in support of nuclear energy in February 1975 were all directly connected with, and highly experienced in the field. Unlike astrophysicist Alfven, the six Physics Nobel Prize winners among them (Alvarez, Bardeen, Bethe, Bloch, Purcell, Rabi, Wigner) have had practical experience with nuclear power, and some of them, like Bethe and Wigner, are among the original developers of nuclear reactors.

When 700 scientists in Alfven's native country Sweden presented the Prime Minister with a statement supporting nuclear power, they did not have to look among the entymologists and psychologists for support. All of the 700 were active in research and technology relating to nuclear power.

The American Nuclear Society has endorsed nuclear power, of course.

Of course?

No, not of course at all. For 21 years, the ANS adamantly refused to endorse nuclear power, because it was more critical of nuclear safety than any of the Johnny-come-latelies. Only in 1975, satisfied at last, the ANS endorsed nuclear power as the safest form of power generation.

So did the 18,000-member Power Engineering Society.[5]

So did the Energy Committee of the 170,000-member Institute of Electrical and Electronics Engineers (IEEE).

So did the 69,000-member Society of Professional Engineers.

So did the National Council of the 39,000-member American Institute of Chemical Engineers.

So did the Board of Directors of the 3,400-member Health Physics Society.

25,000 scientists and engineers signed a "Declaration of Energy Independence" urging increased use of coal and nuclear power and presented it in the White House in 1975, on the second anniversary of the Arab oil embargo. The signers of the petition had a combined total of *two hundred thousand* man-years of experience in electrical power generation.[6]

The "division" among scientists on nuclear power, then, is a peculiar one. To put it brutally, but fairly accurately, it is a division between those who know what they are talking about and those who don't.

THE anti-nuclear activists have a way of getting round this, of course, and a better way than the false pretense (also often used) that a large fraction of scientists is supporting them. The experts, they claim, should disqualify themselves for two reasons: One, the matter is not a technical issue, but a moral one; and two, the experts have a career at stake, therefore their judgement is clouded by a conflict of interests.

Lorna Salzman, an official of the misnamed Friends of the Earth, puts it in the equally misnamed *Bulletin of the Atomic Scientists* as follows:

"No nuclear scientist with a connection to nuclear power profit making has a right to bludgeon government and citizens into accepting a death-dealing technology, a technology that puts private profits and jobs over human health and lives... All scientists having a personal stake in the development of commercial nuclear power should disqualify themselves from the nuclear power discussion and leave the field to citizens who are perfectly capable of determining what endangers them and their freedom."

We shall not meet Lorna Salzman again until the last chapter of this book, by which time I hope to have shown that it is activists like Lorna Salzman who put their unbridled lust for political power over human health and lives, and who do their utmost to brainwash citizens with amateurish sophistry to prevent them from seeing what endangers them and what protects them.

In the meantime, we will note that the conflict of interest and morality arguments are both false and vicious.

The conflict of interest argument is false because it assumes that the career of a nuclear scientist or engineer is limited to commercial power generation. It isn't, of course. Nuclear engineering has a large number of subdisciplines — medical diagnosis and therapy instrumentation, for example — all of which are short of manpower. The argument also overlooks the endorsement by the American Power Society, only a small fraction of whose members are *nuclear* power engineers, and by the American Health Physics Society, which does not live by nuclear power at all, but is, by its statutes, "devoted to the protection of man and his environment from the harmful effects of radiation."

But the argument is also vicious, because it implies that a physician cannot be trusted to cure his patients, for he makes a living only if they are sick. It implies that the police favor crime, for without criminals they would not be needed.

There is no reason to doubt that there are crooks among nuclear engineers, just as there are among nudists, tightrope walkers or diabetics. But the implication that *all* nuclear scientists, or even a considerable fraction, would put their careers over human lives is one that is repugnantly sick.

And so is the "morality" argument. The Friends of the Earth, Ralph Nader, and the other anti-nuclear crusaders have been given the facts often enough; the facts that show non-nuclear methods of power generation to be more dangerous to human health and lives. They have never seriously disputed the point; they have merely ignored it. What kind of morality is it that keeps the public death toll unnecessarily high? What kind of ethics is it that sacrifices human lives?

AMONG nuclear scientists and power engineers, nuclear power is supported by an overwhelming majority. Why, then, is this large number of experts rarely heard?

There are several reasons. First, it has become quite common for nuclear power to be attacked by entymologists, chemists, biologists, sociologists, politicians, journalists and housewives; but only nuclear physicists and engineers are "allowed" to defend it (and even then under the stigma that they have an axe to grind), which in itself tips the scales against them by sheer numbers.

Second, nuclear physicists and engineers, being in the know, have (until recently) laughed off the anti-nuclear hysteria in much the same way as astronomers laugh off believers in a flat earth, or as mathematicians laugh off circle squarers, or as more than a century ago, railroad engineers laughed off the Cassandras who predicted death and pestilence from the railroads. Until very recently, few of them have spoken out in defense of nuclear power, and until the California Initiative campaign was well advanced, there was no organized counter-campaign to combat the anti-nuclear hysteria, certainly none with anywhere near the financial and media support the anti-nuclear crusaders are getting.

Third, and probably most important, when the pro-nuclear side tries to turn the monologue into a debate, it is ignored by the mass media, which give such exaggerated coverage to the blatantly false accusations by Nader and other laymen.

The media bias against nuclear power is evident only to those who have investigated the issue, and remains unknown to millions of TV viewers and magazine readers. The anti-nuclear bias, however, is so strong that in many cases it amounts to censorship — not censorship *of* the press, but censorship *by* the press.

When in January 1975 thirty-four of this country's foremost scientists, including eleven Nobel Prize winners, issued an appeal in favor of nuclear power, the media, in particular the TV networks, ignored it.

"The Republic is in the most serious situation since World War II," said their statement. "Contrary to the scare publicity given to some mistakes that have occurred, no appreciable amount of radioactive material has escaped from any commercial US power reactor... We can see no reasonable alternative to an increased use of nuclear power to satisfy our energy needs..."

Surely the content of the proclamation was newsworthy; the 34 illustrious names signed underneath, if nothing else, made the document unique. But NBC and ABC ignored it; CBS filmed a bit of the briefing by Nobel Laureate Hans Bethe, but never mentioned the statement that had preceded it. Instead, it reported on another of Ralph Nader's warnings against nuclear power, and this was followed by a filmed report which raised questions about the economics, reliability and safety of nuclear power — leaving them unanswered, as though the answers were not known and had not been given a thousand times.

That kind of "fairness" is more damaging, and also more common, than outright censorship as practiced by NBC and ABC on that occasion.

It is well known to professional propagandists that there is one falsehood more vicious than the outright lie: the truth, but not the whole truth. For a drastic illustration, the statement "Mr. Smith probably has not raped any women for the last five weeks, at least not in broad daylight" does not leave Mr. Smith many defenses, for if he protests that it is false, the implication is that he rapes women in broad daylight, too.

This example of the lie by the incomplete truth is, of course, drastic. The networks use this tactic far more subtly, but the underlying trick — the truth, but not the whole truth — is the same.

In February 1975, NBC aired an hour-long "documentary" on nuclear power. It never mentioned the appeal of Nobel laureate Hans Bethe and the other 33 outstanding scientists that had been issued a month earlier. The anti-nuclear opinions were given by Prof. Kendall of the Union of Concerned Scientists. The pro-nuclear point of view was represented by bits and pieces, often only single sentences, cut by an NBC editor from an interview with Dr. Dixie Lee Ray, then chairwoman of the Atomic Energy Commission. Dr Ray is, of course, very knowledgeable about nuclear power, but her name was then unknown to most of the millions of viewers, and the alternation of statements, juxtaposed by the NBC editor, came across as a "debate" between the anti-nuclear scientist and the pro-nuclear bureaucrat.

As usual, the authors of the program were careful to keep the artificial association betwen nuclear power and nuclear bombs alive. The whole program began and ended with a series of nuclear explosions, and the main course was full of them, too: Hiroshima, Nagasaki, Los Alamos, Bikini — the works. One wonders if a documentary on water treatment plants would have been introduced by, interspersed with, and concluded by shots of apartment building fires; after all, in both instances it is a case of oxidation.

After all these explosions, there appears Dr Dixie Lee Ray's face saying "Nuclear explosions in power plants are a physical impossibility." If she explained why this is so, the NBC editor cut it, leaving her statement looking more like an opinion than a scientific fact.

"We are not concerned only with explosions, but also with core melt-downs, which could kill a large number of people," says Kendall's face on the screen in the "reply" selected by the editor.

"The chances of such an accident are one in 10 million years," says the film strip with Dr. Ray's face and voice.

"Maybe so," says John Chancellor, the narrator of the program, "but here is a disaster that did happen, a statistic come alive," and the camera focuses on the wreckage of an H-bomb carrying airplane that crashed in Spain some years ago. "This time, the safeguard worked, and the bomb did not explode, but..."

And so on. It should be noted that not a single lie was told; more than that, NBC can always claim how scrupulously fair they were by letting both sides be heard, if criticism should be raised by some sufficiently powerful person, group or institution.

Yet the exchange above is worse than a lie, for it is the old "incomplete truth" falsehood, leaving the unsuspecting viewer with the impression that probabilities are meaningless numbers, disasters have occurred, nuclear plants are potential A-bombs. As for the millions of viewers who listen to the technical talk with half an ear, but see the Bikini mushroom with both eyes, the idea that nuclear plants are potential nuclear bombs becomes firmly implanted.

And this illustrates only one or two techniques of the electronic brainwashers. Their various techniques have recently been described by Bruce Herschensohn in *The Gods of Antenna.*[7] He identifies 26 techniques (from A to Z), but he seems to have overlooked one of the most effective: Stage a debate by pitting a giant against a pigmy. The technique has proved its worth in politics: Take any issue, and invite a liberal and a conservative to debate it. The liberal is a highly articulate writer, usually handsome; the conservative is all but illiterate, preferably one who has been in politics for the last 50 years. The liberal never gets a chance to annihilate the conservative, because the latter does it all by himself, on account of he don't recall the true fac's. It's all perfectly fair to to both sides, but guess which point of view wins among the viewers?

It works marvelously in politics, and just as well against nuclear power. The young man from Environmental Action scares people with meltdowns, wastes, radioactivity, plutonium, terrorism, and corporate socialism. The old buzzard from the Chamber of Commerce knows even less about plutonium than his opponent, and takes the line that we need newcular energy 'cause of jobs and unemployment and a healthy economy and economic growth, and that's what newcular power is all about. It all looks very fair and balanced, but in effect it asks a simple question: Are you willing to risk thousands of lives so the big corporations do better business? And people will answer like a programmed computer, never suspecting that the question itself is a fraud.

THE anti-nuclear bias is not limited to the networks or even the "liberal" press. It is all too often found in such journals as *Business Week* and even the *Wall Street Journal.*

Business Week does not have a high standard in any respect, least of all in its technical reporting, so that its frequent monumental blunders and biased articles in the field of nuclear energy do not particularly contrast against, say, its exhortations to protect the airlines' anti-competitive nest under the cozy wing of the C.A.B.

However, the *Wall Street Journal* has a reputation of being accurate and publishing corrections on the rare occasions when it does err. That reputation is probably well deserved — but not in the case of nuclear power. The *Wall Street Journal's* articles on nuclear power and related subjects have included statements that were not merely inaccurate, but utterly false; and in none of these cases did the editors correct the errors when they were pointed out to them.

There is, for example the "case of the 23 nuclear power plants," a case of a colossal, but not at all unique, exaggeration. What had actually happened was that in January 1975 a worker at Commonwealth Edison's Dresden Unit 2 in Illinois had discovered, by visual inspection, a hairline crack in a pipe of the emergency cooling system. The crack was so small that it did not leak any moisture. If it had leaked, the automatic monitors would have detected it. And if they hadn't detected it, and the water had leaked out, still nothing would have happened, for the pipe belonged to the back-up cooling system that just stands by in case the primary system should fail. How much radioactivity would have been released? None; the water in this cooling system is as radioactive as the water you drink with your lunch.

In any case, all that did in fact happen was that a worker discovered, by visual inspection, a hairline crack in the pipe. The "incident" (yes, a minute abnormalcy of this type is classified as an "incident" in the nuclear industry) was reported to the Nuclear Regulatory Commission, and what the Commission did is noteworthy for the super-strict standards of nuclear regulation: It ordered all nuclear plants in the country with the same type of pipe shut down for inspection. There were 22 such plants besides Dresden 2, and by April 1975, all but one of them had been inspected after scheduled shut-downs (the 22nd was granted a delay because of a potential power shortage in its area). No cracks were found in any of the remaining 22 plants.

However, that one hairline crack that leaked no water at Dresden 2 became, in the *Wall Street Journal*'s words, "mysterious cracks appearing in the cooling systems of reactors at 23 plants;" and while it is true that it was not the only and by far not the worst misinformer on this occasion (the *New York Times* wrote about failures and consequent shutdowns at 23 plants, for it evidently trusts Nader's hoaxes more than the detailed NRC reports), it once again did not print a correction when the error was pointed out.

It would be nice to think that on this and several other occasions, the *Wall Street Journal* merely misinterpreted something about which its editors know less than nuclear engineers know about stocks and bonds. But that theory must be discarded, for its writers are just as bad when nuclear matters are not linked to physics, but to things as non-technical as union-management disputes, fraud, and even alleged murder.

In a center-page article on plutonium by Burt Schorr, a member of the *Wall Street Journal*'s Washington Bureau, one can read (among several false assertions) about the mysterious fate of an Oklahoma plutonium plant worker, Karen Silkwood, whose body had registered dangerously high plutonium levels, who had been a union activist, who had charged lack of safety provisions in handling plutonium, and who was killed in an auto accident on the very drive she took to meet a newspaper reporter who was investigating plant conditions. The article reports more union charges and is so written as to make the reader suspect foul play. (Nader's "Critical Mass" organization holds memorial ceremonies by candle light for Karen Silkwood.)

Schorr's article does indeed sound ominous — until one examines the rest of the evidence. The Oklahoma state highway patrol reported that an autopsy revealed traces of alcohol and methaqualone (a sedative) in Silkwood's blood, making it most probable that she had dozed off at the wheel (she was killed on the highway by hitting a culvert). But more important in Mr. Schorr's non-technical omissions is his failure to mention the report by a special AEC commission which investigated the union charges of lack of safety. Of the 39 union charges, the commission found only 3 in violation of the tough AEC standards, and the report threw some light on the actions of Karen Silkwood. The monitoring equipment caught her leaving the plant with dangerous amounts of plutonium on two consecutive days, although there had been no accidental release at the plant, and the commission found that her contamination, which included ingested (eaten) plutonium, "probably did not result from an accident or incident within the plant." Moreover, two of the urine samples that she had brought in for inspection and that turned out to be radioactive were proved to have been contaminated *after* they were excreted, so that they must have been doctored by "someone." Though the careful wording of the commission's report does not say so explicitly, the obvious suspicion is that the urine samples were doctored by Silkwood herself.

Schorr's article made no mention of the commission's existence, report or findings. And that is not poor physics; it is poor journalism.

But why pick on the *Wall Street Journal* when there are so many more and much more blatant examples? The reason is in the little word "even." *Even* the *Wall Street Journal,* which on so many other occasions has bucked the fashionable trend, has fallen for Nader's hoaxes, has participated in the anti-nuclear hysteria, and has practiced the new brand of journalism. One does not expect any better from the lesser papers, but when the *Wall Street Journal* writes that plutonium is "a fuel toxic beyond human experience," it is perhaps time to be alarmed.

Not that I suspect its editors, or even Mr. Schorr, of deliberate foul play. What I suspect is that the nuclear opponents have become so vociferous and numerous (among journalists especially) that when the editors are faced by *expert* complaint, they no longer know who is right, and they play it safe. For things are so far gone that the hoaxes are "safer" than the truth.

Few people outside the networks' newsrooms or the editorial offices of the bigger publications know the details of how these hoaxes and distortions get onto the TV screen or into print, and I claim no special knowledge on the point. But there is at least one instance that has been described in detail by an immediate witness who was, at the time, Promotion Director of *Look* magazine. He is Melvin J. Grayson, who together with former *Look* publisher Thomas R. Shepard authored *The Disaster Lobby,*[8] a book on the follies of environmental extremism. It includes a chapter on the role of the media, aptly named "The Closed Fraternity," and among the episodes described in it is one concerning an anti-nuclear propaganda piece authored by *Look* senior editor Jack Shepherd (no relation to author Shepard) in 1970.

The article was called "The Nuclear Threat *Inside* America" and was a string of vicious halftruths — for example, the statement that in 1966, there were 37 accidents at nuclear plants in the US and 6 of them had more than one. What Shepherd did not say was that neither in 1966 or in any other year (up to the present) had there been any reactor-related fatalities in any nuclear power plant. But there may very well have been accidents in 37 plants — a truck backing into the loading dock, for example, or a worker getting his heel caught in a revolving door. None of which is contained in Shepherd's halftruth or three-quarter lie.

When *Look* Promotion Director Grayson saw the stats of Shepherd's article, he was intrigued by one of a long string of allegations: "Some 325 workers [at the AEC Rocky Flats plant near Denver] have been contaminated by radiation since 1953. Fifty-six workers got cancer; 14 have died."

Now Grayson is no expert on nuclear power; in fact, even in the book he seems unaware that Rocky Flats is a weapon plant that has nothing to do with nuclear power. But having had previous experience with Shepherd's brand of journalism, the 14 cancer deaths puzzled him. What exactly was the meaning of that figure in a plant that employed hundreds of workers and had been in operation for 17 years? He did some checking, and found that the cancer death rate at Rocky Flats was no higher than in nudist colonies or stock exchanges, in fact, *it was lower than the cancer death rate for all American adults* (presumably due to the frequent inspections and preventive health care).

Of course, if you will re-read Shepherd's original statement, you will see that he never told a lie, just as he has not raped more than three women in the last two weeks, at least not in broad daylight.

The other allegations were equally truthful, and Grayson pointed out the many halftruths and three-quarter lies in the Shepherd piece to V.C. Myers, President of the *Look* Division of Cowles Communications, suggesting that something be done about it. "Myers," reports Grayson, "exerted all the pressures he could, but to no avail. The editors, with total dominion over the editorial content of the magazine prevailed, and the Shepherd article went to press in its original form."

Grayson reports a number of other hair-raising distortions (not connected with nuclear power) that were published in *Look*, in all cases *after the authors had been presented with the facts*, so that it was not a question of error or incompetence, but of deliberate distortion.

Yet *Look*, says Grayson, was not particularly culpable; "the amount of bias in its editorial department was about average for the period." There was no conspiracy, no meetings on how to slant the facts, no instructions to encourage bias. "What created the bias was the fact that most of *Look*'s editorial people thought pretty much along the same lines and those lines skewed to the left... The men and women who produced *Look* detested big business [and] worshipped the ecological and consumerism reformers..."

Why, then, would a creature like Shepherd, who by Grayson's and Shepard's account wildly distorted the truth, not by error, but by deliberate intent, not simply be fired? "No longer," say the authors of *The Disaster Lobby*, "does the proprietor or board chairman have the last word — or even the first word. Today the hired editors set the policies. They do this under the always effective threat that if management interferes the editors will quit, taking with them the writers, artists, photographers and technicians. And so financially precarious is the publishing business these days that such a threat will cow the most dictatorial of owners."

Shepherd's truth-lie about the cancer deaths at Rocky Flats is typical for statements made in "documentaries" and newscasts of the TV networks, and for the wire services, whose articles and dispatches are blindly reprinted by newspapers all over the country.

In Shepherd's case, we have testimony that the distortions were deliberate, premeditated and printed in spite of the author's better knowledge. How about John Chancellor or Walter Cronkite? Do they really not know that it is impossible for a power plant to blow up in a nuclear explosion? Do they really not know how minuscule the volume of nuclear wastes is or that they can be disposed of more safely than fossil wastes? Do they really not know how much bigger the dangers of fossil fuels are?

Perhaps not. Perhaps they are not deliberate liars. They always have the defense of being grossly incompetent.

AND that, one might think, leaves only the scientific and technical journals to be trusted.

Alas, not necessarily so. Only a very small part of electrical engineering (even of power engineering) is nuclear, and only a very small part of physics (even nuclear physics) is concerned with the generation of power. Only 20,000 of the IEEE's 170,000 members belong to the power engineering group, and only a fraction of these work in *nuclear* power. The remaining groups range from antennas and computers all the way to Engineering English. Similarly, only a small part of nuclear physicists work in nuclear power rather than, say, elementary particle physics, and most physicists work in altogether different branches — optics, astrophysics, solid state physics, etc. These engineers and physicists (let alone biologists, chemists or mathematicians) know little about nuclear power unless they make a special effort to learn about it; yet many of them live in the atmosphere of academia, where the ideologically motivated bias against nuclear power is rife, and the imbalance is reflected in some of the scientific journals. Not in the actual scientific papers; but in the news and comment sections.

Consider a typical case, *Science*.

In the late sixties, when it was the fashion for young radicals to pelt the presidia of various organizations with eggs and tomatoes at convention time, the presidium of the American Association for the Advancement of Science got pelted with eggs and tomatoes at convention time. It was also the fashion to condemn and capitulate; and the AAAS condemned the hooliganism and largely capitulated to it. If the radical punks no longer pelt the AAAS with eggs and tomatoes, it may be because they no longer need to: The AAAS weekly *Science*, in its news and comment columns, has grown societally relevant, involved, legitimate, conscious, concerned, aware, sensitive and progressive.

Nuclear news was until recently covered by Robert Gilette (he has now been awarded a year-long fellowship in journalism at Harvard), whose anti-nuclear bias is ill-concealed. Typically, he reported on the Bethe appeal (p. 21) by briefly mentioning it (*Science* did not print the appeal, though it was short) and then devoting most of the item to what Ralph Nader had to say about it. The whole item was tucked away among other run-of-the-mill news. But when Nader and Kendall delivered their gimmicky petition (p.17) to the White House in August 1975, this non-event rated two thirds of a page in a special box and an 18-point headline over a report by Robert Gilette starting with the halftruth "The American research community is showing new signs of polarization over nuclear power." Gilette also gives figures on the budget of a pro-nuclear Washington lobby, never mentioning that Ralph Nader's anti-nuclear lobby spends some $100,000 a year, and finally this Harvard-fellowship winning journalist characterizes the recently formed Americans for Energy Independence, a group sponsored by nationally known scientists, businessmen, labor leaders and military men, as an organization "whose contributors range from Westinghouse Corporation to a passel of utilities." In all of these halftruths the implication is that the nuclear "debate" is one between scientists and the big corporations.

Note that I am discussing Gilette's journalism, not his scientific competence. The latter was revealed when he misinterpreted an AEC report on serious nuclear accidents with release of radioactivity to the public and wrote "The report indicates that [by 1980] one such accident each year may become a virtual certainty."[9]

Since Gilette left for Harvard, his brand of journalism has been ably continued by P.M. Boffey. *Science* has given space to the portrait of Ed Koupal, the sponsor of the California Nuclear Initiative, who says "The only physics I ever had was Ex-Lax,"[10] and when Daniel Ford, the non-scientist from the Union of Concerned Scientists, complained that the *Scientific American* had refused to let him and Kendall attack an article by Nobel Prize winner Hans Bethe, *Science* spread this totally unjustified complaint under an 18-point headline spreading across two pages,[11] a strange practice for a journal that has censored pro-nuclear news, such as a proclamation by 11 Nobel Prize winners.

Gilette's pseudo-journalism is, of course, more damaging than that of the gutter press, since it carries the prestige of being printed in a scientific journal. Whether this prestige is still justified is debatable; where nuclear disasters or the glorification of Ed Koupal is concerned, *The National Enquirer* and *Hustler* have shown more reticence.*

I WILL take good care to separate the comments on the *Bulletin of Atomic Scientists* by a new paragraph, for it is a purely ideological-political publication that attempts to masquerade as a scientific journal. Masquerade and deception are, indeed, its hallmarks, for it is not only published under a totally misleading title (it has long been edited by a non-scientist and atomic scientists usually write in it only to refute its alarmist science fiction), but it also carries a proud list of sponsors including Albert Einstein, Hans Bethe, A.H. Compton, Leo Szilard, and other famous scientists. These did, indeed, sponsor the journal in 1945; but most of them no longer live, and those that do, either write there to dissent (Hans Bethe) or have themselves forsaken science for politics (Linus Pauling). The most prolific contributor is one D.D. Comey, a kremlinologist who masquerades as a scientist and, as head of an outfit called "Businessmen for the Public Interest," as a businessman as well. His amateurish sophistry, when refuted by reputable scientists, is given the last word in a rebuttal; and when he himself attacks a scientist, he is given the last word again.

The flavor of this "scientific" journal is perhaps best captured by looking at the contents of a single issue (February 1976): "Secrecy and

* I have to eat my words here, for since I wrote this, the *National Enquirer* (2/17/1976) ran a page-wide headline *The Horrifying Day a Blazing A-Plant Threatened 11 Million Americans* with subtitle *People in 9 States Only Minutes from Death.* The hair-raising horror story underneath was transparently based on D.D. Comey's version of the Browns Ferry fire (see Chapter 3).

Security" (Editorial); "Covert Action: Swampland of American Foreign Policy" by Sen. F. Church (Insert, in a colored box, a poem called "National Anathema:" *Oh C.I.A. can you see / By the Chile dawn light / How profoundly you failed / In your late great scheming...* etc.); "The Week We Almost Went to War" (an article claiming that the Cuban missile crisis was unnecessary and provoked by the US); "The dangerous drift in uranium enrichment;" a four-page advertisement to join and contribute money to the Continental Walk for unilateral nuclear disarmament; and so forth up to the inevitable Comey who argues the public risk from uranium tailings over the next 80,000 years.

And this is the journal quoted incessantly by the nuclear critics with the implication that the quote has scientific authority!

It is also the journal from whose subscribers Ralph Nader selected names to which to send his petition against nuclear power, and then boasted of "2,300 scientists" who signed it.

Lorna Salzman's quotations in this and other chapters are, of course, taken from that journal, too.

THE foregoing observations were intended to show that there is no nuclear debate, only a monologue by anti-nuclear laymen. But the truth is that in an objective and dispassionate atmosphere there could not be much of a debate, at least not between people who share some fundamental values such as the sanctity of human life and the need to minimize the health hazards in an industrial society.

For nuclear power is not abortion, inflation, crime abatement or minority rights, where the problems — let alone the solutions — are fuzzy, ill-defined, unquantifiable and truly controversial. The problems of nuclear power are sharply defined, quantifiable, measurable and well understood. There are viable solutions to all of them. When the experts debate a problem — such as nuclear waste disposal — the reason for the debate is not the absence of a solution, but the large number of satisfactory alternatives.

But to make that point, one does not have to compare nuclear power with crime abatement. It is sufficient to compare it with fossil-fired power. Strange as it may seem, far more is known about the health hazards of nuclear power than about those associated with fossil-burning plants. For example, we do not know the exact extent to which some diseases are caused by air pollution, we do not know the exact relative contributions to air pollution by industry and the automobile, we do not know how to stop air pollution by fossil-burning plants entirely (except to go nuclear), and we do not even know how to accurately measure all of the pollutants emitted by the stack of a fossil burning plant. Recent experiments on the blood of blood donors in large cities, analyzed on weekdays and Sundays, have even challenged the "obvious" assumption that the automobile is the major contributor to photochemicals in the atmosphere, and that natural vegetation is not a major contributor to hydrocarbon pollution; and while this is not yet established, the controversy shows how fundamental the gaps in our knowledge about air pollution are.

Not so with nuclear power. The basic hazard is the release of radioactivity, and the effects of radiation on the human body are unusually well understood. Unlike chemical pollutants, which cause cancer and a multitude of other diseases, radioactivity by accidental release from a nuclear plant can cause only two diseases — cancer and radiation sickness (genetic mutations are so improbable that they are omitted in this brief introduction), and unlike the case of pollutants by fossil-burning plants, the relationship between exposure to radioactivity and the incidence of these diseases is firmly established — and not merely by qualitative criteria, but by quantitative relations in hard numbers.

Given these hard numbers of the risks to human health and to the environment associated with nuclear power, and comparing them to the somewhat fuzzier, but still unambiguously high, numbers of the risks associated with other forms of energy conversion, nuclear power emerges convincingly as the safest.

Moreover, nuclear power emerges not as the safest in *some* aspects, but the safest in *all* aspects, not excluding terrorism and sabotage, and certainly not excluding major accidents and waste disposal.

That is why there could not be very much of a debate if there were a debate and not a monologue.

However, in the din of the nuclear monologue, the pro-nuclear peeps are difficult enough to hear; but even so, not too many of the peeps have been concerned with the comparison of risks. More often they have compared the risks with the benefits.

That may have been a mistake. If large sections of the population oppose nuclear power, it cannot be *entirely* due to Shepherd and his colleagues in the Closed Fraternity. Of course the Shepherds have it easy. People do not suspect bakers of putting poison in their bread, because they have a rough idea of how bread is baked; but when it comes to neutrons, isotopes, plutonium and fission products, they have a fear of The Unknown and it is easy for the Shepherds to plant absurd associations between nuclear bombs and nuclear power in their minds. Yet people have grown to accept very dangerous things, such as gasoline refining and air traffic control, about which most of them do not know much, either. And a considerable campaign against the fluoridization of water did not manage to scare them. So something else must be wrong.

It may well be that what is wrong is the nuclear advocates' tactic of weighing the risks of nuclear power against its benefits. There is nothing wrong with that in itself, for such analyses have been made frequently and nuclear power has always come out with flying colors. But these analyses have not managed to reassure everybody.

And no wonder. To say that "a major nuclear disaster every 10 million years is worth the benefits of electric power" is, first and foremost, not a matter of fact, but a matter of opinion, a subjective judgement of values. It is not at all quantifiable and it is highly debatable, especially by one who does not know what else is involved.

To say that "Everybody, consciously or not, weighs risks and benefits every time he gets into a car or an aircraft" is doubtlessly true, but it leaves room for many Yes, but's.

"Yes, but I take these risks voluntarily, nuclear power is forced on me."

"Yes, but an air crash involves only tens of fatalities, a nuclear disaster would involve thousands."

It so happens that both of these sample objections are erroneous. [They *are* erroneous: When a woman is about to have a baby, how much choice does she *really* have (in America) whether to walk to the hospital or to go by some gasoline-powered vehicle? And as for aircraft crashes, almost every year there is one with more than 100 fatalities; nuclear disasters *might* entail tens with a smaller probability, and they have not incurred a single one yet.]

Nevertheless, the *Yes, but*'s will continue. "Yes, but if you wait long enough, a disaster killing thousands will eventually happen." (True, but not very impressive: If you wait long enough, a Pulitzer-Prize winning one-eyed twin will be stamped to death by a female elephant.) And so it goes, back and forth, on and on.

While it is my opinion that the benefits of nuclear power far outweigh its risks, this book will not press the point and only very occasionally touch on it. It has been made too often, and with less than complete success.

Instead of comparing the risks of nuclear power to its benefits, this book will compare the risks of nuclear power to the risks of any other form of energy conversion.

The statement "Per billion megawatt-hours of generated electricity, generated by the corresponding fuel, either 1036 coal miners, but only 20 uranium miners lose their lives"[12] is not a statement of opinion or a subjective judgement of values, but an assertion whose truth can be checked out by universally accepted methods; it deals with hard, measurable, verifiable and comparable numbers. I have no doubt that Ralph Nader has a "yes, but" in retort, but I do not believe it can be very convincing, especially when other, similar statements of risk comparison are considered.

Such a comparison has at least two advantages: First, it compares quantities of the same dimensions. In comparing risks to benefits, one sooner or later runs into the question of how many dollars a human life is worth. In comparing risks only, we shall compare probabilities to probabilities, deaths to deaths, injuries to injuries, and disease incidences to disease incidences. We shall never have to compare apples with oranges.

Second, the comparison will rid us of some irrelevant problems, such as energy conservation. Nuclear power is unnecessary, claims Nader, because we don't need any more energy if we cut our demand sufficiently. The argument is false, but we need not go into it, for besides being false, it is also irrelevant. Indeed, suppose that it were feasible to cut US energy consumption by 50% (which I do not for a moment believe); shouldn't we make sure that the remaining 50% are supplied by the safest possible method?

2

Some Basics

Plutonium was named for Pluto, the god of hell. It is arguably the most toxic substance known.
Dr Elise Jerard, chairperson of the Independent Phi Beta Kappa Environmental Study Group.

When things went awry at the Enrico Fermi reactor near Detroit, four million people went about their business in happy ignorance, while technicians gingerly tinkered with the renegade's invisible interior. They knew what the public did not — a mistake could trigger a nuclear explosion.
M.E. Gale in the *New York Times Book Review*, 30 November 1975.

The real dangers of a nuclear power plant arise when something goes wrong, and just like medical students must learn anatomy before pathology, so we must take a quick look at a healthy nuclear plant.

Electricity is, most often, produced by moving a conductor in a magnetic field, and that is essentially what is going on in an electric generator; for our purposes it is sufficient to think of an electric generator as something that produces electricity when its shaft is turned. The turning is done by a turbine on the same shaft; the whole arrangement is known as a turbogenerator.

The turbine, if it is a hydraulic turbine, is turned by water rushing through it from a reservoir behind a dam. But ony about 12% of the US electric capacity is hydroelectric, and the fraction is growing smaller as the total capacity grows and the US is running out of sites to dam rivers.

The great majority of power plants are thermal plants: Their turbines are either steam or gas turbines. Steam is, of course, itself a

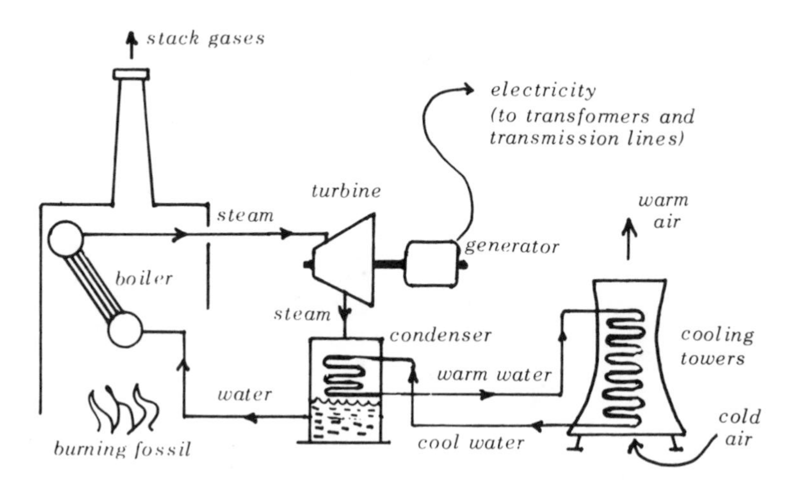

How a power plant works. Schematic drawing showing the barest principles, omitting economizers, superheaters, reheaters, feedwater heaters, high and low pressure turbines, pumps and other elements.

gas, but it's too late to change the terminology now. A gas turbine runs on the hot gases resulting directly from the combustion of fuel; it is very similar to the engine of a jet plane. Gas turbines, like everything else, have vices and virtues. The main vices are low efficiency and costly fuel; the main virtue is the rapidity with which they can be put into and out of operation. They are therefore usually used only during the peak hours to tide the plant over the period of greatest demand.

The machines that run day and night to supply the "base load" are steam turbines; they are far and away the most important for generating electric power. Water is heated and turned into steam; the superheated steam (up to 550°C or about 1,000°F) is let loose into the turbines, where it presses on the blades and makes the turbine, with the attached electric generator, spin. The steam is not only pressed into the turbine, it is also sucked out of it, for it passes from the turbine into a condenser, where it is liquefied back into water to return to the beginning of the cycle. The same water (liquid or steam) runs through the boiler, turbine, condenser, back to the boiler all the time.

The condenser is a set of tubes kept cool so that the steam condenses into water on their surface. The condenser tubes are cooled by a separate circuit of cooling water.

What happens to the cooling water after it has extracted heat from the condenser tubes and turned warm is really none of our business, for we are out to examine the health hazards of going nuclear; yet having come this far, it is as well to remark on two popular misconceptions.

Sometimes the cooling water is taken from a nearby river and returned to it after it has been heated by passing through the condensers. This may warm up the river by a few degrees near the plant and has been given the ludicrous name of "thermal pollution." The warmer water does indeed drive out some species of aquatic life, but it provides a habitat for other species for which the water had previously been too cold. The warmer water does not kill aquatic life, it merely shifts the spectrum of species, and very slightly at that (see Chapter 5).

Contrary to popular belief, a nuclear plant does not produce significantly more waste heat than a fossil-fired plant of the same capacity: The most efficient nuclear reactors perform to within two percentage points of the most efficient fossil-fired plants, and the average reactor performs to within two percentage points of the efficiency of the average fossil-fired plant (Chapter 5).

"Thermal pollution" is mostly an abuse of language; it has as much pollution as the East European "people's democracies" have democracy.

Another and more common way of cooling the cooling water (and preserving most of it to go back to work) is to let it drip through a cooling tower, where it gives up its heat to the air drawn through the cooling tower by natural draft or fans. Cooling towers are most often high concrete structures, often with a white plume of "smoke" coming out of it; on cold days with high relative humidity that plume rises high into the sky and can look quite ominous, causing the local instant ecologists to protest indignantly about air pollution.

But the plume is merely condensed water vapor, the same stuff that clouds in the sky are made of. The pollutants of a fossil-fired power plant do not come out of the cooling towers, but out of the stack (a nuclear plant doesn't have one); and the deadliest pollutants are quite invisible.

To complete this rough description of a power plant, there remains the first link in the chain: the heat that turns water into steam. In a conventional plant, this is provided by burning a fossil fuel — coal, oil or gas — under the boiler; the hot combustion gases heat the water tubes of the boiler, though much of the heat is wasted and escapes through the stack.

In a nuclear plant, the heat turning water into steam is generated by a nuclear reactor; the steam then turns the turbogenerators in the same way as fossil-fired plants. And that is the first thing to note about nuclear power: Strictly speaking, there is no such thing as nuclear electric power (at present, anyway); there is only nuclear heat, and that heat is used to generate electricity in the same way as in fossil-burning plants. There is no difference in the electricity supplied to the consumer; it may have been produced by either type of plant, and most often it is produced by both — by both types of plant working into a common grid that distributes power to consumers.

There is a nuclear plant now under construction in West Germany that will use hot helium, heated in the reactor, to drive the turbogenerators; but otherwise the world's commercial power plants using nuclear power use steam to run the turbines. However, even in that German plant the electricity produced will be the same as any other.

The heat produced in nuclear reactors can, of course, be used for other purposes than generating electric power. It is used, for example, for propelling ships and submarines (giving them great ranges of operation without refueling), and there are plans for nuclear steel furnaces, for example. However, we shall be concerned only with the issue of generating electric power commercially for civilian use.

To see how a nuclear reactor generates heat, we have to review some elementary physics.

The smallest entity of a chemically homogenous substance — distilled water, say, or pure kitchen salt — that still has the same physical and chemical properties as the substance in bulk is called a molecule. In all but a handful of cases, a molecule is a combination of atoms, of which there are 92 different kinds in nature; they are the atoms of the elements. The lightest atom is that of hydrogen, the heaviest (among the natural atoms) is uranium. Beyond uranium there are the "transuranic" elements that do not normally occur in nature, but can be artificially produced; one of them is the element plutonium.

Each atom consists of a positively charged nucleus and a shell of negatively charged electrons. In most artists' conceptions, from comic strips to trade marks, the atom is depicted as a little solar system with the nucleus in the middle and the electrons whirling round it along planet-like orbits. That is not an altogether correct model of the atom, but it will do for our purposes.

The nucleus of an atom is very much heavier than its electron shells — many thousands of times heavier. However, the negative electric charge of the electron shell, in a stable atom, just equals the positive charge of the nucleus, so that the two charges cancel and the atom is electrically neutral.

For example, an oxygen molecule consists of two oxygen atoms; each oxygen atom has 8 electrons whirling round the nucleus whose positive charge equals, except for the opposite sign, the 8 negative electron charges, so that the entire atom is electrically neutral.

To "split the atom" is easy; it is far more difficult to split the nucleus. Indeed, to split the atom, it is sufficient to tear out one or two electrons from the shell, and this is done quite easily (for example, in a fluorescent light). To split the nucleus is a different story.

There was a time when people thought that the nucleus consisted only of protons, particles having the same charge as an electron, only positive, but much heavier (about 2,000 times heavier). But in 1932, the existence of another particle was demonstrated: the neutron, which is just as heavy as a proton, but has no electric charge. The presence of a neutron in the nucleus will thus make the atom heavier, but will not disturb the electrical balance between the nucleus and the electron shell.

It would have been a simple world if it had remained at that. But as time went on, more and more nuclear particles were discovered, and more are being discovered all the time. The "nuclear zoo" is now so large and bewildering that it has theoretical physicists deeply disturbed, and many people believe that a new, fundamental and revolutionary discovery is needed to put some genuine order into this monumental puzzle.

Fortunately, the nuclear zoo has little or nothing to do with nuclear power, and for our purposes it will be quite alright to pretend that there are only three types of particles in an atom: protons, neutrons and electrons. The negatively charged electrons whirl round the nucleus, which contains at least one positively charged proton and possibly one or more electrically neutral neutrons.

The number of electrons, which is also the number of protons in a neutral atom, is called the atomic number. Hydrogen, with one proton in the nucleus and a single electron orbiting it, has atomic number 1; uranium has 92 of each, and its atomic number is therefore 92. It is the atomic number that determines the chemical (and most of the physical) properties of an element. If we could simply double the number of protons and electrons in a hydrogen atom, we would have not hydrogen, but helium — a gas with utterly different properties. If it were possible to cut iron atoms (atomic number 26) in half in such a way that the new atoms had 13 protons and electrons each, the new atoms would not be iron, but aluminum (atomic number 13).

The atomic number is unaffected by the neutrons that may be present in the nucleus; so what do *they* do? Chemically, nothing. A hydrogen atom with a neutron in its nucleus (in addition to the regular proton) still combines with oxygen to form water, or with chlorine to form hydrochloric acid. What the neutrons do is change some of the nuclear properties of the atom. Obviously, they increase the weight of the nucleus; the "mass number" of the atom, which is the number of protons plus neutrons, or simply the weight of the nucleus in multiples of one proton, will increase by one with every additional neutron. (The atomic mass number is not quite the same thing as atomic weight, but let's not walk off into *that* side street.)

As far as we are concerned here, the mass number does nothing except to provide us with a convenient way of classifying the different types of chemically equivalent atoms. "Different types of chemically equivalent atoms" can be expressed by a single word: *isotopes*. Isotopes of an element are atoms with the same atomic number, and therefore with the same chemical properties, but with different mass numbers. Isotopes of the same element have the same number of protons, but different numbers of neutrons in their nuclei.

For example, the isotopes of hydrogen (atomic number 1) are hydrogen proper (1 proton, mass number 1), deuterium (1 proton + 1 neutron, mass number 2), and tritium (1 proton + 2 neutrons, mass number 3). Hydrogen isotopes have their own distinct names, the only ones to be so honored. The isotopes of the other elements do not have distinct names; their mass number is simply added to the chemical symbol. For example, the isotopes of carbon (C) are C 12, C 13, C 14. Since the atomic number of carbon is 6, carbon 14 has 6 protons plus 8 neutrons in its nucleus.

Different isotopes of the same element are chemically indistinguishable, because chemical effects are associated with the electron shell. The difference affects only nuclear properties, in particular, whether the atom is radioactive or stable, and whether it is fissile or not. (We will use *fissile* and *fissionable* interchangeably, though professional jargon makes a slight difference between them.)

Radioactivity is a subject that will interest us shortly, especially its effects on the human body. At this point it is enough to say that it amounts to radiation of different types that is emitted when a nucleus disintegrates or decays. In some elements, such as uranium or radium, the decay is natural and proceeds without man-made prodding. In other cases, radioactivity may be artificially induced, most often by shooting a neutron into the nucleus and causing it to fall apart.

The latter case is of interest for self-sustained fission of nuclei. As the nucleus splits into two or more fragments, it may also emit one or more of its neutrons. If these neutrons are absorbed by other nuclei, they will split in turn and emit more neutrons for further splitting of other nuclei. A multiplicative chain reaction will take place. And each time a nucleus splits, energy is released — at least in the nuclei at the heavy end of the table of elements. The released energy was previously pent up in binding the nucleus together.

There are only four types of nuclei that are thus fissionable, and only one of them occurs naturally in significant quantities: uranium 235, i.e., the isotope of uranium with mass number 235 (92 protons plus 143 neutrons in the nucleus). The average number of neutrons available for causing further fissions of a U 235 nucleus lies between 2 and 3, depending on the energy of the neutrons. (The actual number in an individual fission is, of course, an integer; the *mean* number is 2.07, and the fractional part is due to averaging over the various possibilities in a fission.)

Of course, in practice the two neutrons generated in a fission do not both cause another fission in U 235, even if it were possible to get a quantity of perfectly pure U 235. A lot of other things besides absorption by another U 235 nucleus can happen to a neutron; for example, it can simply fly out of the volume containing the uranium into the adjacent air or other medium.

This is a good place to discuss the A-bomb, made of U 235 or plutonium; not because it has anything to do with nuclear power, but to see why a nuclear explosion cannot possibly take place in a power plant.

In a uranium bomb, the material is purified ("enriched"), at great cost, to more than 90% U 235. In natural uranium ore, almost all of the uranium is U 238, which is not fissionable. The uranium contains only about 0.7% of the fissile isotope U 235, and this small fraction must first be increased, i.e., the uranium enriched, to 90%.

Even so, in a small amount of 90% enriched uranium, there will be no chain reaction, because most of the neutrons shooting out of the nuclei after each fission will simply leave the uranium instead of hitting other uranium nuclei. To get a multiplicative chain reaction, the average number of neutrons absorbed by other nuclei per fission of a nucleus must be greater than one. There must, therefore, be enough of the material, in the form of a sphere (largest volume to surface ratio), to ensure that most of the liberated neutrons will be absorbed by other nuclei before they can escape from the volume. If the sphere can be held together long enough (about one millionth of a second) for the chain reaction to take place, rather than blow itself out by tearing the sphere apart again after it has started, a nuclear explosion will take place. In a bomb, this is achieved by joining two or more pieces of uranium (or plutonium), all of subcritical size, into a sphere of more than critical mass. They are "joined" by firing them against each other by high-explosives in a gun-like mechanism.

Since there are less than 3 neutrons per fission available to cause further fissions, and the non-fissile material will absorb some of these, it follows that such a bomb would quite certainly fail to explode if the material contained an insufficient concentration of the U 235 isotope, no matter how large the sphere were made. In practice, the uranium in nuclear weapons contains more than 90% of U 235, i.e., it is enriched to more than 90%.

But the uranium fuel used in power plants contains not 90%, but only 3.5% of the fissile isotope U 235 (the rest is mostly U 238), which is way below even the theoretical minimum, and it is therefore utterly impossible for it to undergo an explosive nuclear chain reaction.

As a matter of fact, there are other reasons why a nuclear explosion is utterly impossible in a power plant, but we will not go into them. After all, if a man has lost both his legs, he cannot run; do we have to go through all the other reasons why he cannot qualify for the olympic team?

And yet a nuclear power reactor also works by a chain reaction — the neutrons generated by one fission cause more fissions — so if the reactor somehow gets out of hand, couldn't there be a nuclear explosion after all?

No, there couldn't. The chain reaction in a reactor is quite different from that in a bomb. Even in a runaway reactor, 100 fissions would cause no more than 101 fissions in the next generation. More important, the time elapsing between the fissions of one generation and the resulting next generation, even in a runaway reactor, is about 100,000 times longer than in a nuclear explosion. That is a difference greater than that between a murderous flood and a drizzle.

So there is no getting around it — inducing a nuclear explosion in 3% enriched uranium is absolutely impossible — it would violate physical laws.

It is equally impossible to induce a nuclear explosion in the plutonium oxide mixed with uranium oxide that will be used as "mixed oxide" reactor fuel as soon as reprocessing of spent fuel is legalized. The "wrong material" inhibition is no longer valid in this case (though plutonium bombs are made of pure plutonium, and of a different isotope mix at that), but the geometric configuration alone, not to mention many other reasons, makes a nuclear explosion, even in a runaway reactor, impossible. To say that a nuclear explosion in a mixed-oxide-fueled reactor is more likely than in a uranium-fueled one makes as much sense as to say that hot water is more likely to catch fire than cold water.

And with that, good-bye to bombs; they have very little to do with nuclear power.

FROM the world that is large enough to see, we are used to the idea that the harder we hit something, the more likely it is to break. It is the high impact velocity of a hammer that will crack a nut. But nuclei are not nuts, and they live by different rules. A U 235 nucleus is more likely to absorb a slow neutron than a fast one, and it has to absorb one before it will split. So if we must have an analogy, a neutron is more like a political agitator who sets out to divide a community. Walking from door to door, he may be able to polarize the community by his arguments; he is less likely to achieve this by shouting a few slogans as he quickly drives through the town in a loudspeaker van. And most of the time he will have no effect either way.

So it is with neutrons, and to make them more effective for splitting the uranium nuclei, only 3% of which are fissionable anyway (the U 235 isotopes), they must first be slowed down from the high velocities with which they were emitted in the fissions that generated them. This is done by a material called a *moderator*, which does not absorb the neutrons, but bumps them back into the uranium with a slower velocity. Among the materials that make good moderators are hydrocarbons, beryllium, carbon, but above all, water (or rather the hydrogen in water, for neutron reflection is a nuclear, not a chemical, process).

After the bump, two things happen: The neutron is slowed down by it, which was the main idea, and the particle that got bumped in the moderator is now itself set into motion. But motion of a particle — up to a molecule — is simply heat: The difference between hot water and cold water is simply that the molecules of the former are wiggling around much faster than those in the cold water. The moderator, then, will heat up under the constant bombardment of the neutrons coming out of the uranium (and even more by the heat from the fuel rods). Some of these neutrons will be reflected back into the uranium, and having given up some of their energy, they now move more slowly and are more likely to be captured by a U 235 nucleus. If that causes that nucleus to fission (it doesn't always), the results are, among other fission products, fast neutrons ready to undergo the same treatment. The other fission products, too, collide with whatever they run into (moderator, fuel, other fission products), again turning their kinetic or motional energy into heat. That is essentially how nuclear energy (the energy needed to bind nuclear particles together into a common nucleus) is converted into heat by fission.

The process may or may not continue: It can die out, or it can just be self-sustaining, or it can grow more intensive. Just like a nation grows, remains at a steady population or dies out depending on its average fertility rate (number of children born per 1,000 women of child-bearing age), so the process of fission depends on the number of neutrons produced by one generation of fissions and capable to produce the next generation. If the number of neutrons produced in the next generation is the same as in the last, the process is self-sustaining and heat will be produced at a steady rate; this state is called critical. If the next generation produces less neutrons than the last, the process will gradually die out, and the state is called subcritical; an increase of neutron generation, and therefore of heat production, corresponds to a super-critical state.

The rate of heat production can thus be controlled by controlling the number of neutrons that are allowed to beget the next generation of fissions, and this can be done quite easily by material that absorbs neutrons (and does not fission). There are many materials that have this capability, for example, boron, cadmium, or hafnium.

By inserting such neutron-absorbing material between the uranium and the moderator, the neutrons coming out of the uranium, instead of being slowed down, will be taken out of circulation altogether. In practice, the neutron absorbing material is in the form of control rods, which can be moved so as to let more or less neutrons from the fuel reach the moderator. This provides a continuous range of power output, from total shutdown to capacity power. For any position of the control rods, the process is entirely stable and predictable.

NOW let us take a brief look at how these physical principles are applied in engineering practice. The fuel — uranium oxide — has the form of pellets, which are inserted into stainless steel or zirconium tubes. The resulting "fuel rods" are mounted vertically and parallel to each other to form a precise pattern in the reactor core. The control

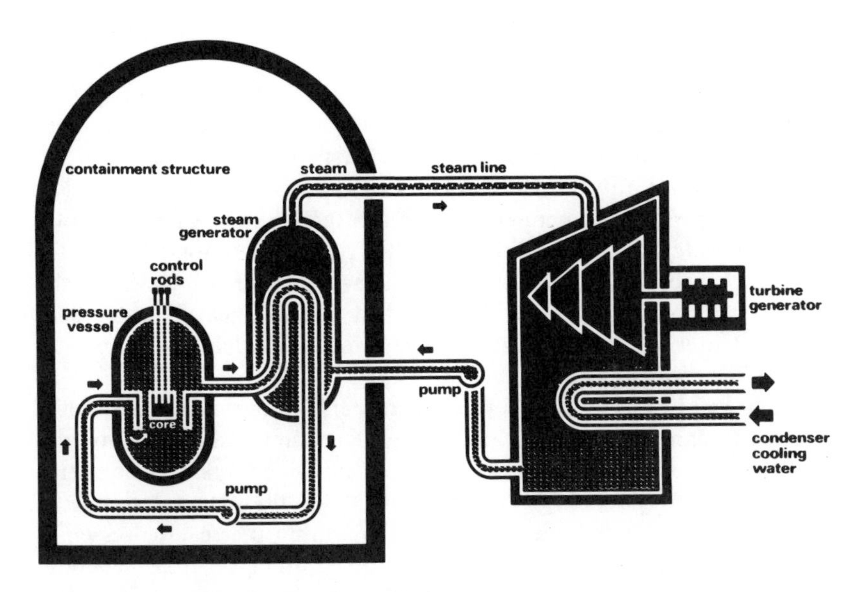

Pressurized water reactor (PWR). The water in the pressure vessel circuit is kept under pressure so that it does not turn into steam at high temperatures. In a boiling water reactor (BWR) there is no heat exchanger (marked "steam generator" above); the steam from the pressure vessel goes directly into the turbine, and the water from the condenser goes directly back into the pressure vessel.

rods can be slid in or out between the fuel rods; the remaining space is filled by the moderator, and the entire core is surrounded by the reactor vessel, about 20 feet in diameter and 45 feet high, weighing several hundred tons.

There are many types of reactors, but the two types most commonly used in the US at present are the Boiling Water Reactor (BWR) and Pressurized Water Reactor (PWR). Both use water as a moderator, and both use water for a double purpose: moderator and coolant. The water surrounding the fuel heats up, slowly or quickly depending on how far the control rods are pulled out of the core, and the hot water generates steam to drive the turbogenerators.

In a boiling water reactor, the water in the core is brought to a boil and its steam feeds the turbines directly. In a pressurized water reactor, the water is kept under pressure in a closed circuit, so that it cannot turn into steam. It transfers its heat in a heat exchanger (steam generator) to another water circuit, where the secondary water is turned into steam to drive the turbines.

Once the steam has been produced, the generation of electricity is the same as in a fossil-fired power plant.

There are other types of reactors, e.g., the High-Temperature Gas Reactor, which uses a gas (helium) as a coolant; it is both more efficient and safer than light water reactors, and it could be the reactor of the distant future. However, as yet there have been only two such reactors in commercial operation in the US, and we shall err on the safe side by ignoring this and other types of reactors. Except for breeder reactors, they all work on the same basic principles as the light water reactors. (Light water is just water; "heavy" water contains deuterium rather than hydrogen and is used in the Canadian CANDU reactor, but not in the US.)

The brief description above concerns only those components of a nuclear plant that are essential to its operation. Added to this are the numerous components that serve to enhance the safety of operation. The one that first hits the eye is the containment building, a huge, massive, dome-like structure surrounding the reactor vessel. It is made of four-feet thick, heavily reinforced and steel-lined concrete. It would protect the outside world from radioactivity released by a reactor after a melt-down; but it also protects the reactor from the much more likely disasters that threaten the reactor from the outside world, such as hurricanes, tornadoes, earthquakes and aircraft crashes. The tough AEC (now NRC) regulations require the containment building not only to withstand gales of more than 180 miles per hour, but even jetliners crashing into it at landing speeds.

Incidentally, the material and structure of the containment building is not unlike the U-boat pens built by the Germans on the French coast during World War II. In spite of savage round-the-clock bombing and the use of special "blockbuster" bombs, the Allies failed to crack them. This is a thought that comes to mind in view of the latest scare by the nuclear critics — what if civil war comes to America? It is very doubtful that the AEC had this possibility in mind when it drew up the regulations, but it so happens that even in that case nuclear reactors would be safer than most other places — certainly much safer than the neighborhood of large storage facilities of oil and natural gas, for example.

There is only one type of major accident that can happen at a nuclear plant *qua* nuclear plant, and that is the release of radioactivity. A steam turbine could, of course, be accidentally damaged, or for that matter, the mail truck could run over the security guard at the gate, but these are not accidents peculiar to a nuclear plant. The

Five steel workers carefully tie together 6 layers of thick reinforcing bars to form a web over the steel-structured containment building. Over this will go a layer of concrete 3½ feet thick; the mesh is so tight that special vibrators have to be used to force the concrete mix through it. The containment building, which is required to withstand the crash of a jetliner at landing speed, is the last line of defense (not counting favorable weather conditions) against a major nuclear accident. If a meltdown does occur, it will contain the released radioactivity with a probability close to one.

The unfinished area in the center will be completed after installation of the reactor and other equipment. (*AIF archives.*)

real danger is release of radioactivity — nuclear explosion, as we have seen, is pure horror fiction. The most dangerous accident is LOCA — Loss of Coolant Accident, which may (or may not) be followed by a melt-down of the core, which in turn may (or may not) be followed by the release of radioactivity out of the containment building, and that may (or may not) result in deaths among the general public. Even then, however, the corpses would not pile up in the streets as depicted in the sick fantasies of the alarmists, for these deaths would be long in coming — weeks or more for radiation sickness, and 10 to 45 years for cancer.

However, we leave these unpleasant details for later, when we shall compare them with the far more unpleasant details of major accidents associated with fossil-burning plants. In this brief survey of a healthy nuclear plant we will only metion that the reactor vessel — itself a steel vessel with walls between 6 and 11 inches thick and weighing 450 tons — has welded into it not only pipes providing the entrance and exits of the coolant (water), but also the pipes of the ECCS — the Emergency Core Cooling System. This is a system with independent pipes, independent water, independent pumps, and even (if the need arises) with an independent power supply, which forces the cooling water into the reactor vessel if for some reason the regular coolant should be lost — for example, by a pipe of the water circulation system bursting. The ECCS is activated automatically by monitors watching over the state of the reactor, but should the automatic system fail, there are provisions for manual control.

If this were a book on nuclear safety (rather than a book comparing nuclear hazards to alternative hazards), there would follow a long list of the many other safety measures, safety equipment, safety regulations, and a description of the safety philosophy on which all of these are based — defense in depth, redundancy of components, and many more, leaving a minimum to human decisions and providing safeguards if human error does occur. But this is not a book on nuclear safety, and so we will forego this list, which has been enumerated very often, and with all too little effect on the public.

To give just a little idea of what is involved, we will only take a quick look at how the control rods shut down the reactor automatically if something goes wrong. It is a very unimportant example in itself, but it illustrates the philosophy on which safety measures are based, a philosophy that is also applied in ordinary passenger elevators.

Roughly 120,000 Americans die in accidents every year. The no. 1 killer, of course, are motor vehicle accidents, with some 50,000 deaths; the no. 2 killer is less well known — accidental falls, which in some years kill as many as 20,000. The remainder are aircraft and railroad accidents, poisoning, fire, electrocution, explosions, firearms, blows by falling objects, snake bites, and a lot of others.

Some of these deaths are freakish — people freeze to death by getting accidentally locked into refrigerators, or they are killed by an aircraft crashing into their house. But there is one type of accident remarkably absent from the list — people plunging to their deaths in passenger elevators. Why? Only a small fraction of Americans ever handle firearms, yet some 3,000 are killed annually by accidental discharges (not counting homicides and suicides). But everybody, at some time, has used an elevator, and millions use one several times a day. Why aren't people plunging to their deaths by the thousands?

Because of the regular inspections enforced by law? Hardly; if it were that simple, there would be no failures of critical components in automobiles. The reason is that every elevator has a gadget that prevents it from plunging down the shaft — powerful jaws that grip the guide rails of the cabin and stop it from dropping. And why does this gadget hardly ever fail to work? Because it is not *activated* when something goes *wrong*, but it is *kept inoperative* only if everything is *right*. For example, the jaws are directly connected to the cable on which the cabin is suspended, and the tension of the cable keeps the jaws inoperative. If the cable should snap, or lose its tension for some other reason, the jaws return to their "natural" position, jamming the cabin in the guide rails long before it can gather additional speed.

It is this philosophy, the philosophy of "don't activate safety measures when something goes wrong, keep them inoperative only if everything works right" is widely applied in nuclear reactors. The control rods, for example, are (usually) vertical and pulled up to increase power, so that their "natural" position is the shut-off position, to which they will drop back under their own weight. What prevents them from dropping back are electromagnets powered directly by the electrical output of the plant. If for some unknown reason the plant suddenly stops generating electric power, the magnets let go, and the control rods instantly shut off the neutron flow.

This is just one example of one of the principles on which the safety measures of a nuclear plant are based. In the few cases when a sudden shut-down was needed, this particular device has worked well.

For example, when in October 1966 a metal plate broke loose in the Fermi I reactor, it partially blocked the flow of coolant to two out of 100 fuel assemblies, so that these two overheated and some of their fuel melted. There was no difficulty in promptly shutting down the reactor, and all safety systems worked exactly as planned. The reactor was later repaired and resumed operation.

You may not believe this (and it *is* hard to believe), but this incident is the subject of the book *We Almost Lost Detroit.*[1] Its flagrantly false story has been coolly shattered in an expert report.[2]

Fermi I was an experimental fast breeder reactor, which uses sodium rather than water as the working fluid, and therefore its safeguards (such as emergency core cooling) were somewhat different from those of a conventional light water reactor.

But among the reasons why we didn't lose Detroit is one that applies to all reactors, conventional or not: the principle of defense in depth. If the reactor had lost its coolant, it would have been automatically replaced. And if it hadn't, the containment building would have contained the radioactivity. And if it hadn't (though it is hard to see why not), it would have dispersed into the atmosphere without doing any harm. And if it hadn't, because a temperature inversion kept it near the ground, a slight wind in an unfortunate direction would have had to blow it 30 miles to Detroit before a Detroit fly got hurt.

Or so it would seem. In the case of Fermi I, as we shall see in Chapter 3, a Detroit fly *could not* have been hurt even in the worst case, which makes this silly book even more despicable. "An unforgiving technology," says the introduction, implying that nuclear power has no room for human error. In fact, *only* nuclear power, with its defense in depth, has considerable room for human error.

This was borne out by another incident that has been turned into a scare, the Browns Ferry fire in March 1975. An incredible chain of human errors piled up in that case, beginning with a workman inspecting electrical cables, in the last quarter of the 20th century, with a candle. Yet no radioactivity escaped, and would not have escaped if still more errors had been piled on the chain, for not even the first line of defense was ever broken, or even close to being broken.

But what would have happened if some shnook had inspected an oil refinery or a liquid natural gas tank with a candle? He wouldn't be there to tell us, for other energy facilities do not have a defense in depth. There would have been an explosion of the type that happens all the time, and that's it. There is no room for error in working with large quantities of oil, gasoline or gas, and when they release their energy, they often *do* (not *might*) kill more people than a nuclear accident, even if one does happen, is likely to do.

THERE is one more basic concept that we should briefly visit, and that is the concept of energy itself.

Energy is the capacity to do work. Work and energy are two quite similar concepts (measured in the same units); in fact, the difference between them is simply the algebraic sign — plus or minus, just like the difference between credit and debit or assets and liabilities. So what is work?

Perhaps the simplest type of work is mechanical work, which equals force times distance. Weight, for example, is a force, the force of gravity, and if a man lifts 5 lbs of potatoes 3 feet from the ground, his work done against gravity is 15 foot-pounds.

The foot-pound is not the only unit of energy; there are others such as the BTU, calorie, erg, joule, kilowatt-hour, electron-volt, and more. Since they all measure the same thing, energy, they can be mutually converted, e.g., 1 calorie = 4.18 joules; but why are there so many units? For many reasons, some of them historical; but the main reason is convenience. A British Thermal Unit (BTU) is the heat required to raise 1 pound of water by 1°F, which makes it convenient for heat transfer engineering, but highly inept for a nuclear physicist, who prefers to use the electron-volt — the energy necessary to move the charge of one electron through a potential difference of 1 volt.

We will not let ourselves be confused by all these units. We shall use only one — the kilowatt-hour (kWh), and anybody who prefers joules, BTU's or anything else need only look up a table of conversion factors. For example, 1 kWh = 3,600,000 joules = 3,410 BTU.

To get a feel for the size of a kWh, we first have to go to the watt, which is not a unit of energy, but of *power*. Lifting the same weight to the same height always results in the same energy, no matter whether it was done fast or slowly. The concept that takes into account the speed with which it was done is power; power is the *rate* of doing work (expending energy). Power is the work performed *per unit time.*

If the work of 15 ft-lbs (lifting 5 lbs of potatoes 3 feet high) is performed by a forklift in 1 second, the useful power developed by the forklift is 15 foot-pounds per second, whereas if the same work is performed by a small child who lifts the potatoes one by one and takes 100 seconds to accomplish the job, then the (average) power is only 0.15 foot-pounds per second.

There are again many units of power, e.g., the horsepower, which James Watt thought corresponded to the power developed by a horse. Actually, no horse can do work at the rate of 1 HP for very long, and many horses are not able to reach 1 HP even for a short time. However, we shall only use the *watt*, or more often its multiples *kilowatt* (1,000 watts) and *megawatt* (1 million watts).

The precise definition of the watt need not worry us here, and we will only give a few examples to get a feel for its size. One watt is very little power; for example, a light consuming only 1 W can be seen in the dark, but it is not strong enough to illuminate anything. The most commonly used light bulb in the home consumes 60 W (of which only about 5% are turned into light, the rest is heat). An electric pressing iron consumes about 1 kW, and a clothes drier about 3 kW. An electric kitchen range, with the oven and 4 heating elements turned on will consume about 6 kW. Industry, of course, uses power on a much larger scale. The motors blowing air through a wind tunnel where aircraft models are tested can consume several *mega*watts, the equivalent of a thousand clothes driers or kitchen ranges.

It follows that even for a small residential community a power plant producing several hundreds of kilowatts is needed. For a city, the plant will have to generate several megawatts; as a very rough rule of thumb, a city of a million will need a plant of 1,000 MW capacity. Capacity power is the maximum power the plant is capable of generating. It will be reached or approached only during the peak hours (time of peak demand). The average power supplied during the whole day will, of course, be considerably smaller.

The *base load*, or demand that is there throughout the day, is supplied by base plants, the big, reliable plants that pollute little (pollution standards are set by the quantity per day). As the load increases, more generators are brought on line, for example, gas-turbine driven ones. The peak load generators, which work only during part of the day, and some of them only for an hour or two, usually have smaller power; they also tend to be less efficient, less reliable and more polluting. Why? Well, if a utility has some excellent and some poor equipment, which will it use for the base load, and which will it switch in only under duress?

Nuclear plants now range from several hundreds to several thousands of megawatts. They are invariably used to supply the base load. Any utility that has a "nuke" will bring in the inferior stuff only as needed.

AND finally, let us briefly look into radioactivity and its effects on human health.

A good way to start is to note that radioactivity is a perfectly natural phenomenon. The ground we walk on is radioactive; so is our blood; so is the food we eat; so is the air we breathe.

Perhaps you say, ah, but that is all negligibly small radioactivity compared with what a nuclear plant puts into the environment.

Wrong. It is the nuclear plants that are negligible. The background radiation for the average US citizen amounts to some 250 millirems per year, of which more than half is due to natural sources, and the remainder is mostly due to medical equipment. What nuclear plants add to these 250 millirems is a piddling 0.003 mrems, a fly sneezing into the wind.

But let's start at the beginning. Radioactivity is the radiation released in the disintegration, or decay, of an atomic nucleus. It spontaneously breaks apart and shoots out particles, and this is what constitutes radioactivity; the remaining heavy fragments (if any) are new atoms, which may be either stable or themselves radioactive, i.e., liable to disintegrate at some future time, in which case they are called "daughters" of the original substance. Naturally radioactive elements go through several generations of such daughters and daughters' daughters; they all eventually end up as lead, which is stable.

There are four types of radioactive radiation — alpha, beta, gamma, and neutrons, depending on what type of particle is shot out from the nucleus in its disintegration. Alpha particles are helium nuclei (2 protons plus 2 neutrons), beta particles are electrons, and a gamma particle is a short burst of electromagnetic radiation — a photon of high energy, or a quantum of light (invisible because of its extremely short wavelength).

Only gamma rays and neutrons are capable of penetrating matter to any significant depth; it takes many feet of earth or several feet of concrete to reduce their intensity to almost nothing. On the other hand, beta particles can travel only a few feet in air before they are absorbed, and alpha particles only a few inches; a sheet of paper will absorb them.

A radioactive element can radiate any or all of these types of radiation, but a given isotope always radiates the same type or types. For example, plutonium is primarily an alpha emitter, which means that it is not at all dangerous at a distance, and even close to it, as little as a newspaper will act as a shield against its radiation. Plutonium is highly dangerous when inhaled (and moderately toxic when eaten or absorbed through the skin), though even then to call it "the most toxic substance known to man" is melodramatic nonsense, as we shall see.

The timing of when a particular nucleus will disintegrate is entirely random and impossible to predict, though the probability laws governing its behavior are exactly known. Given a certain amount of atoms of a substance, it is, for example, exactly known how long it will take for half of the original amount to disintegrate. This time is called the *halflife* of the corresponding radioactive isotope. After a halflife, half of the originally present atoms are still intact in the form of the original substance; another halflife will leave a half of that half, or a quarter; a third halflife will leave one eighth; and so forth.

Different isotopes have different halflives. Polonium 213 has a halflife of only 4 millionths of a second, but uranium 238 has a halflife of 4.5 billion years (which is fortunate, or none would now be left).

It stands to reason that for a given amount of atoms, the intensity of the radioactivity will be the smaller the longer the halflife — just as the same amount of water in a reservoir can give rise to a short and deadly flood, or to a prolonged and gentle flow if it is let out slowly. Yet this point does not seem to have impressed many environmentalists, whose lamentations grow the more shrill and nervous the longer the halflife of some radioactive isotope. They have never considered what the halflife of stable substances (such as arsenic or cyanide) is — it is infinite.

When a person is exposed to radioactive radiation, the amount received, i.e., the intensity of the radiation multiplied by the exposure time, is called the delivered *dose*. Doses are measured in roentgens, which are based on the total delivered energy. However, the biological damage to tissue does not depend on the energy alone (which is minute, anyway); a unit that takes into account the relative

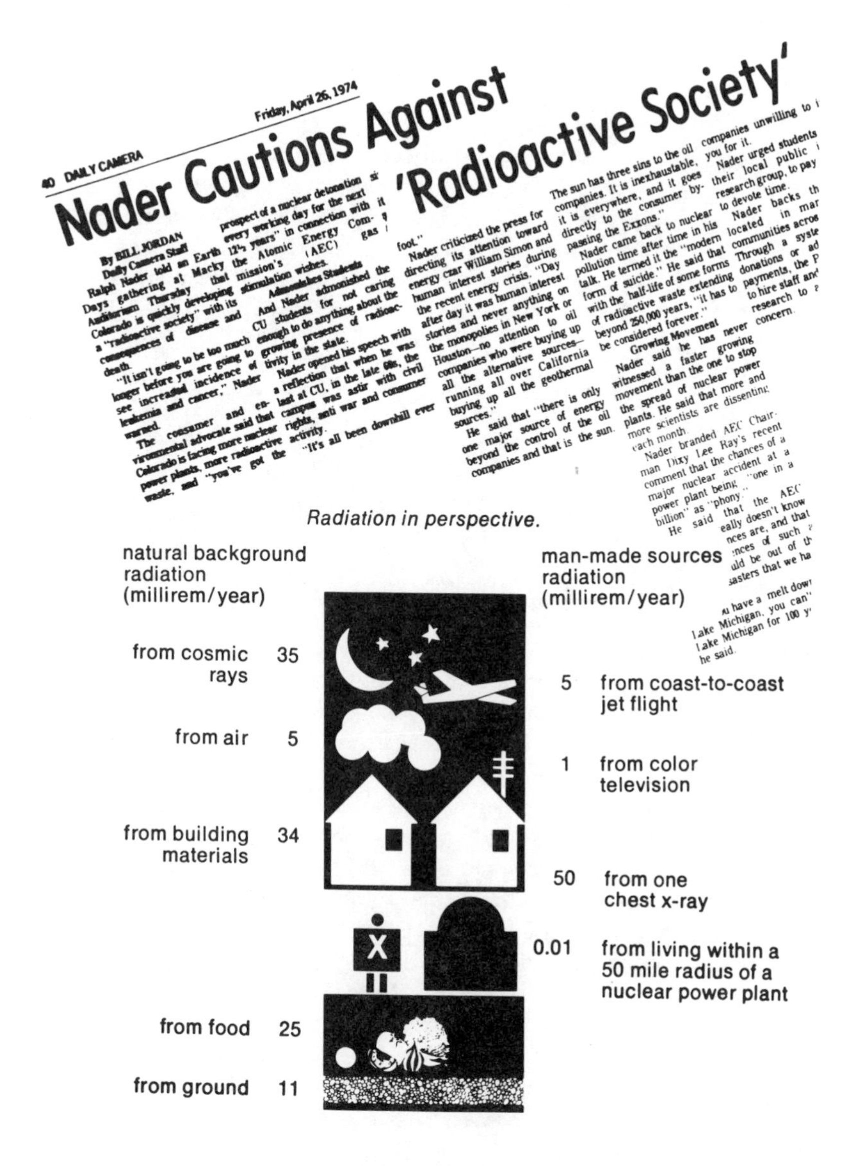

40 DAILY CAMERA — Friday, April 26, 1974

Nader Cautions Against 'Radioactive Society'

By BILL JORDAN
Daily Camera Staff

Ralph Nader told an Earth Days gathering at Macky Auditorium Thursday that Colorado is quickly developing a "radioactive society" with its consequences of disease and death.

"It isn't going to be too much longer before you are going to see increased incidence of leukemia and cancer," Nader warned.

The consumer and environmental advocate said that Colorado is facing more nuclear power plants, more radioactive waste, and "you've got the prospect of a nuclear detonation every working day for the next 12½ years" in connection with the Atomic Energy Commission's (AEC) gas stimulation wishes.

Admonishes Students

And Nader admonished the CU students for not caring enough to do anything about the growing presence of radioactivity in the state.

Nader opened his speech with a reflection that when he was last at CU, in the late 60s, the campus was astir with civil rights, anti war and consumer activity.

"It's all been downhill ever

foot."

Nader criticized the press for directing its attention toward energy czar William Simon and human interest stories during the recent energy crisis. "Day after day it was human interest stories and never anything on the monopolies in New York or Houston—no attention to oil companies who were buying up all the alternative sources—running all over California buying up all the geothermal sources."

He said that "there is only one major source of energy beyond the control of the oil companies and that is the sun. The sun has three sins to the oil companies. It is inexhaustible, it is everywhere, and it goes directly to the consumer bypassing the Exxons."

Nader came back to nuclear pollution time after time in his talk. He termed it the "modern form of suicide." He said that with the half-life of some forms of radioactive waste extending beyond 250,000 years, "it has to be considered forever."

Growing Movement

Nader said he has never witnessed a faster growing movement than the one to stop the spread of nuclear power plants. He said that more and more scientists are dissenting each month.

Nader branded AEC Chairman Dixy Lee Ray's recent comment that the chances of a major nuclear accident at a power plant being "one in a billion" as "phony."

He said that the AEC ...eally doesn't know ...nces are, and that ...nces of such ...uld be out of th... ...sasters that we ha...

...u have a melt dow... Lake Michigan, you can'... Lake Michigan for 100 y... he said.

companies unwilling to ... you for it.

Nader urged students ... their local public ... research group, to pay ... to devote time.

Nader backs th... located in mar... communities acros... Through a syste... donations or ad... payments, the P... to hire staff an... research to ... concern.

Radiation in perspective.

Like a witch-doctor frightening savages in the jungle with a transistor radio, Nader frightens his flock of adulators with radioactivity ("It isn't going to be too much longer before you are going to see increased incidence of cancer and leukemia"). On this occasion, he spoke at the University of Colorado; and the flight from Washington and back gave him more radioactivity than a nuclear plant would give him in a lifetime.

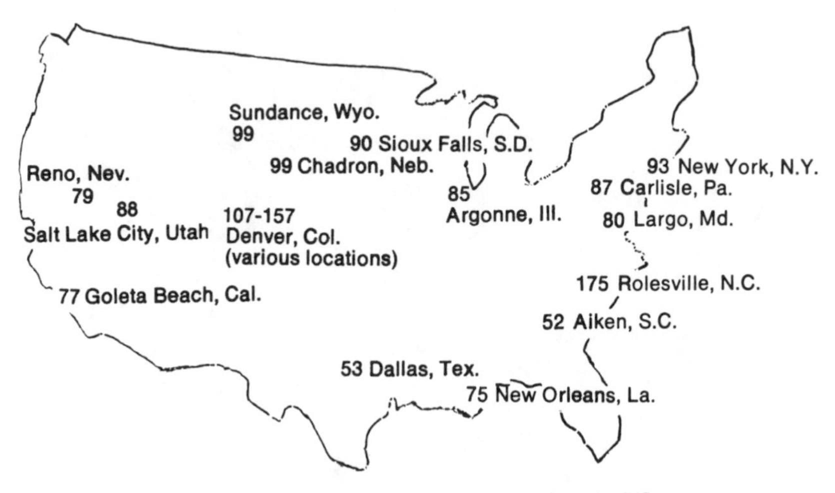

Natural background radiation (terrestrial and cosmic) in the US as measured in 1965. The numbers are in millirems/year. A nuclear power plant adds 0.01 mrem/year.

biological effectiveness of different types of radiation is the *roentgen equivalent man* or *rem*. One thousandth of a rem is a millirem (mrem).

To get an idea of the size of a millirem, consider the following numbers. The International Commission on Radiological Protection has set 500 mrems as the maximum permissible annual dose that an individual should receive. The figure is conservative, i.e., on the safe side, as is the case with all such standards. There are areas in India and Brazil with monazite sands, rich in thorium and uranium, which give the population an average dose of 1,500 mrems/year, or three times the international permissible standard; yet studies of these populations have revealed no unusual effects.[3]

A single chest X-ray will expose the patient to some 50 mrems; a coast-to-coast jet flight will expose the passengers to some 5 additional mrems; watching color television will deliver an average of 1 mrem/year. Yet all of these doses together are smaller than the dose the average US resident obtains from Mother Nature: 130 mrems/year. Most of this comes from cosmic rays, the ground, and from building materials.

Cosmic rays are gamma rays reaching us from outer space. The atmosphere acts as a partial shield against them, so that at high altitudes. such as Colorado or Wyoming, this component is much stronger than at sea level, where the cosmic ray component delivers

Bumper sticker sold and promoted by Environmental Action of Colorado. By merely living in Colorado, its inhabitants get between 30 to 100 mrems/year more than they would get at sea level; the difference amounts to what would be produced by more than 10,000 nuclear plants (all within 50 miles of the "victim").

an annual dose of about 35 mrem; this dose roughly doubles for every mile of altitude. Altitude is also the main reason, though not the only one, why the natural background radiation varies from place to place. For example, in Denver, Colo., the annual dose from the natural background radiation amounts to 157 mrems, whereas in Aiken, S.C., it amounts only to 52 mrems.

The next largest component usually comes from building materials, such as granite. Grand Central Station in Manhattan could not be licensed as a nuclear reactor, because the radiation from its granite blocks would violate NRC standards. The ground delivers a dose of about 10 mrems/year, and 5 mrems/year come from the air.

Food is radioactive, too, delivering an average of 25 mrems/year to the US citizen. The food intake results in a normal potassium 40 level in the blood, and this delivers an "internal" dose (not counted in the 130 mrems/year of the natural background) of no less than 20 mrem/year. A small part of this radiation is also radiated out of the body, so that when the anti-nuclear fanatics hold a meeting, they receive much more radiation from each other than they would from a nuclear reactor. Or, to quote Dr. Edward Teller, "In sleeping with a woman, one gets just slightly less radioactivity than from a nuclear reactor; but to sleep with two women is very, very dangerous."

To the 130 mrem/year of natural background radiation one must add the dose received from man-made equipment. This amounts, for the average US resident, to about 120 mrems/year, bringing the total to about 250 mrem/year. By far the greatest part of the man-made radiation is due to X-ray diagnostics (103 mrem/year), and most of the remainder also goes for medical equipment (therapeutics 6, radiopharmaceutical 2). Global fall-out, i.e., rain that brings down radioactive isotopes released into the atmosphere by human activity (formerly mostly nuclear bomb tests, now mostly industrial releases) amounts to about 4 mrem/year. Color television watching results in an average exposure of 1 mrem/year, and the remainder are little sources that produce almost nothing, such as the luminescent dial of a watch and — yes, and nuclear plants.

How much do all the US nuclear plants add to the dose of 250 mrem/year that the average US citizen receives already? About 0.003 mrem/year. Yes, that is what the nuclear critics are protesting: 0.003 mrems on top of the 250 mrems that they get anyway.[4]

Both numbers are averages. In individual cases, people can get more; a Coloradan gets more natural background radiation and an immediate neighbor of a nuclear plant gets more from the routine emissions of a nuclear plant. NRC regulations limit the maximum exposure for the latter to 5 mrem/year, though in fact, these emissions are very small and not near this upper limit. But even if a person did get a full 5 mrems in some year from a nuclear plant, his added chance of dying from cancer would be the same as those due to smoking *a single cigarette during that year.* On the other hand, Coloradans are exposed to an additional 35 mrem/year merely by living at a higher altitude; and unlike the 5 mrems at the boundary of a nuclear plant, which are rarely reached and even then absorbed by only a few people, *all* Coloradans get their additional 35 mrems *every* year.

Even so, should people living at the boundary of a nuclear plant move away to reduce their risk? Prof. Bernard L. Cohen of the University of Pittsburgh, a past Chairman of the American Physical Society Nuclear Division, has an answer to that. If they do, he says, they shouldn't move so that they must drive to work by an additional 0.1 mile a day — the statistics will be against them, for they then take a bigger risk of being killed in an auto accident.[5]

WHAT exactly are the health hazards of excessive doses of radioactivity, such as might come about by a major nuclear accident?

There are three such possible effects: radiation sickness, cancer and genetic mutations. Please note that in discussing them we will switch from millirems to full rems, units 1,000 times larger.

Radiation sickness results from exposures exceeding 100 rems (or 20,000 times more than is permitted to accumulate in the course of one year at the boundaries of a nuclear power plant). It is due to the induced malfunction of the bone marrow producing white blood corpuscles, and victims die (if they do die) in a time from a few days to a few weeks after exposure. However, if death does not result (400 rems is roughly the level where half of the exposed victims die), the patient recovers within a few weeks and all symptoms disappear. Radiation sickness is what would cause the "early deaths" in a major nuclear accident; the "delayed deaths" would be due to latent cancers. A dose of more than 100 rems is an improbably strong one, and deaths by radiation sickness among the public could occur only under an extremely unlikely combination of circumstances leading to the contamination of a populated area by large quantities of concentrated, high-level radioactive material (Chapter 3). The more important threat is that of cancer.

Genetic defects in offspring due to radiation exposure of the parents are a well known effect produced in experiments with animals; it has, however, never been observed in humans, not even in Hiroshima and Nagasaki, in spite of extremely thorough and intensive investigations.[6] The reason, very probably, is not that such effects do not exist — when it comes to health effects, man is not so very different from other mammals — but rather that they are too small to be observed. Genetic effects due to spontaneous mutations in the sex cells occur widely in any population even without man-made radioactivity; in the US, no less than 3% of all live births exhibit such effects, ranging from an extra finger or toe to more serious defects such as diseases that show up later in life. It is obviously very difficult to detect any additional effects against this large background even with detailed information and sophisticated statistical methods. In any case, whatever the reason, no genetic effects in humans have ever been observed as a result of radioactivity, and not for lack of trying.

This leaves cancer as the most important health hazard due to excessive doses of radioactivity. Contrary to popular misconceptions, this is a subject which is unusually well understood, no doubt due to the intensive and massive research carried out on the subject during the last 40 years by prestigious institutions all over the world. (One experiment at the Oak Ridge National Laboratory involved the irradiation and subsequent microscopic examination of each of 50,000 mice.[7])

Apart from the detailed research on general laws and tendencies, there is also a substantial amount of empirical data on the effect of excessive doses on humans. There are the 24,000 Japanese who were exposed to an average of about 130 rems in the two bomb explosions in 1945, and among whom more than 100 excess cancer deaths occurred. ("Excess" means the cancer deaths beyond those that would have occurred without exposure; due to the large numbers involved, the statistical estimate of these "normal" deaths is very accurate.)

15,000 people in Great Britain were exposed to heavy doses of X-rays (almost 400 rems) in treating arthritis of the spine before the danger of X-rays was fully known; this again resulted in more than 100 excess cancer deaths. In the US, thousands of miners (mostly uranium miners) inhaled radon, a radioactive gas, and some of them received doses to the lung approaching 5,000 rems. Between 1915 and 1935, there were 775 American women employed in painting radium numerals on watch dials; they used to lick the brushes to point them, and there were similar situations in other countries. Painstaking research of the records in all of these and similar situations has established a relation between radiation dose and excess cancer incidence which is not seriously disputed by any one.[8]

Not everyone who receives a large dose of radiation dies of, or even contracts, cancer. For a given dose, a certain fraction of the exposed population will contract it, or more to the point, a certain fraction of those who would not have contracted it otherwise will contract it. It is therefore usual to express the hazard due to radiation by the corresponding increase in probability of dying of cancer. The "normal" probability of dying of cancer for the average American now stands at 16.8%; this probability is increased by 0.018% for every rem of radiation absorbed by his body. This is the figure based on the 1972 report by the Committee on Biological Effects of Ionizing Radiation (BEIR) of the US National Academy of Sciences and the National Research Council. It is also in agreement with the figures published by the United Nations Scientific Committee on Effects of Atomic Radiation (UNSCEAR); both institutions are under surveillance of standard-setting institutions such as the International Commission on Radiological Protection (ICRP).

The reader is again reminded that our previous discussion of natural background radiation, permissible radiation of nuclear power plants, etc., expressed the radiation in *milli*rems, or multiples of one thousandth of one rem. Now that we are discussing the incidence of cancer, however, we are using the rem, a unit one thousand times bigger. Even so, each rem, or 1,000 millirems, of exposure adds only very little to the already high risk of contracting cancer due to non-nuclear causes; in fact, in America, it increases the chance of dying of cancer from 16.8% to 16.818%. This is true if the entire body absorbs one rem of radiation; if the radiation is concentrated onto some organ (for example, the lung, due to radioactive particles that have been inhaled), the risk may be higher, and the interested reader is referred to the BEIR report for details, but even so, the increased risk per rem is very small. Only doses of tens or hundreds of rems of radiation will appreciably increase one's chances of contracting cancer beyond what these chances are already.

These are the naked facts, and they are not mentioned here to play down the danger of radioactivity. Strong radioactivity can kill, and has killed, large numbers of people, though none as yet have lost their lives at a commercial nuclear plant. Besides, there is no need to play down the potential cancers due to radioactive releases. As we shall see, the cancers due to fossil-fired plants are not potential cancers; they are cancers that kill here and now, and they don't kill abstract figures in computer programs, they kill people who were still alive yesterday.

IN ALL of the foregoing we have used the so-called "linear hypothesis" that the incidence of cancer increases linearly with the radiation dose, i.e., if the dose is increased by some factor, the probability of contracting cancer will increase by the same factor. This hypothesis agrees well with observation in the region where most data are available, around 0.1 to 100 rems. But for very low levels, we have only one point where we are certain of the result, the point zero: No radiation at all produces no radiation-induced cancers at all.

What about very small doses, like 1 millirem? This is the dose most people get by merely living on earth for a couple of days. How is one to measure such a small amount against the background of a bewildering number of other factors?

Although direct measurement is difficult or impossible, the indirect evidence suggests strongly that there is a "threshold" value of radiation below which it is harmless. There are at least three indicators supporting the existence of a threshold: first, the evidence from experiments on animals under strictly controlled conditions; second, the well known fact that tissue slightly damaged by radiation will heal if it has the time — which it does with low doses; and third, radiotherapy on cancerous tissue would not work if there were no threshold, for the radiation damages both healthy and cancerous, rapidly multiplying cells.[9]

Nuclear critics have spent much time disputing the existence of a threshold. For example, Gofman and Tamplin, in their grotesquely biased book *Poisoned Power,* devote much space to arguing (but presenting no evidence) against it.

But the argument is not merely in all probability wrong, it is, above all, irrelevant. For the radiation protection standards and risk calculations — in the US, at any rate — use the linear hypothesis all the way down to zero, that is, just as if the deniers of a threshold value were right. As is customary in US regulation, errors are always introduced on the safe side. Other standard-setting institutions, e.g., UNSCEAR, do not accept the linear hypothesis for low radiation levels, though they have not come up with an alternative, either. The US National Council on Radiation Protection, which sets the standards for maximum permissible exposure, accepts the linear hypothesis, knowing it is erring on the safe side, and acceptance of the hypothesis makes the issue of the existence of a threshold utterly irrelevant.

3

Major Accidents

[A nuclear accident would result in] up to 100,000 deaths and the destruction of an area the size of Pennsylvania

Ralph Nader, address to the Joint Session of the Massachusetts Legislature, March 21, 1974.

Owing to the widespread anti-nuclear propaganda, not many people realize that nuclear power is safer than other methods of generating electric power. But even those who do realize it often oppose nuclear power, or are reluctant to endorse it, because they fear a major nuclear accident.

"No matter how small the probability of a major accident," they say, "it is not impossible; and if one does happen, the consequences will be so terrible that they should not be risked at all."

That view is once again based on misinformation, for the risk of a major accident is far higher in the coal, oil and gas cycles (and for hydropower) than it is in the nuclear cycle. *Not only are major accidents with fossil fuels and hydropower far more probable, but their consequences can be more terrible, too.*

The various scenarios of a nuclear accident leading to a significant number of deaths invariably involves a chain of independent events, each of which has a minute probability. If the same thing were done for fossil fuels, one would have to consider the number of lives lost in a hypothetical accident in which a jetliner crashes into an oil storage complex whose explosion wipes out a nearby city. There is, however, no need to engage into such fantasies, for simple events such as explosions of oil or gas storage tanks have already cost hundreds of lives and could easily lead to single accidents with a death toll in the tens of thousands.

CONSIDER first a major nuclear accident. A nuclear explosion in a reactor is impossible, as we have seen (pp. 41-42). The one and only major threat to the public is a release of large amounts of radioactivity within a short time. There is virtually only one type of accident that can lead to such a disaster — the Loss of Coolant Accident (LOCA). It could happen in the light-water reactors now in use, though even this possibility is greatly reduced in High-Temperature Gas Reactors which, in spite of the financial bind of its producing company, may yet prove to be the reactors of the future.

A loss-of-coolant accident threatens if the water which absorbs the heat from the fuel rods (p. 45) should leak out. The vessel itself, weighing several hundred tons. made of steel several inches thick and subjected to thorough tests before going into operation, could not develop such a leak, but the pipes carrying water to and from the vessel could, even though they are continuously monitored for leaks and designed to withstand earthquakes. The safety measures assume not merely a simple leak such as might result from a small crack, but a "guillotine cut" in which the pipe is cut clean through and the two ends are severed from each other so as to allow the water to gush out without impediment.

If the water were to leak out, and no safety measures were taken to replace it, the control rods, as in all other malfunctions, would drop back under their own weight and shut off the chain reaction in the uranium instantly. However, heat due to the continuing radioactivity in the fission products — the spent fuel in the fuel rods — would continue to be generated, and if no countermeasures were taken, the temperature of the fuel rods could rise to the melting point of the cladding of the fuel rods.

To eliminate this threat, every nuclear power reactor has an Emergency Core Cooling System (ECCS) with independent pipes, pumps and water to pump cooling water into the core if for some reason the normally present water should begin flowing out.

The LOFT (Loss of Fluid Test) facility in Idaho is designed to provide accurate tests of the effectiveness of emergency core cooling during a large simulated loss-of-coolant accident. The first of a series of tests, in March 1976, was a complete success.

The ECCS is designed for some wildly pessimistic assumptions, such as the "guillotine cut" just mentioned. It is required to go into operation instantly and automatically on the ocurrence of a leak, but can also be operated manually if the automatic activation should fail.

There has as yet been no sudden leak in a commercial reactor (certainly not at the Browns Ferry fire in 1975), and the nuclear critics claim that no one knows whether the ECCS would work. The AEC therefore built a multi-million test facility in Idaho for an actual test — the equivalent of sinking a ship to see if the lifeboats would work. But already the critics are claiming that the facility is too small and that the simulation will involve much less power to be extinguished than in the usual 1,000 MW nuclear reactor — the equivalent of saying "You will never know whether the lifeboats work if you just sink a small freighter — you have to sink the Queen

Elizabeth II to be sure." When the tests are carried out in 1976, they will win both ways. If they succeed, they will say the test facility was too small (at 55 MW of heat!); if it fails they will say "We told you so."*

All of which are incidental considerations, for what would happen if there were a loss-of-coolant accident *and* the ECCS failed to replace the lost coolant? In that case, most people think, disaster and massive loss of life would follow.

Not so. The Rasmussen report puts the number of lost lives due to loss-of-coolant accidents, *if there is one*, at an average of less than one. In most cases the dollar damage would be large, but no lives would be lost among the public, and very probably none even inside the plant. What would happen is that the fission products inside the fuel rods would (unlike the uranium, which is easily controlled) produce heat without significant cooling until the metal cladding of the fuel rods melted, and a red-hot mass of metal and fission products would eventually flow down and start melting the steel pressure vessel surrounding the core. This is several inches thick and weighs several hundred tons, so that it takes time to melt through — a marked difference from other accidents, such as explosions or air crashes, in which there is no time for warnings and countermeasures.

In any case, it is not the red-hot and radioactive goo melting through the pressure vessel and flowing down onto the concrete floor that would be dangerous. That would melt through the concrete into the ground, where it would dissipate its heat, and from where it could be removed without major complications.

Dr. R.P. Hammond, a widely respected nuclear scientist with more than 30 years experience, says "If I had to contend with such material — and I have had some first hand experience in cleaning up radioactive spills — I cannot think of a place where I would prefer to have it than far underground. It would be completely shielded by the overlying earth and concrete, it would be enclosed in a thick pocket of fused earth... At a radius of 20 feet or so

---

* The first of a series of tests was performed on March 4, 1976, and was successful. "The successful initial test involved a simulated rupture in the pressurized primary cooling system... The water that suddenly began to be lost was promptly replenished by the emergency core cooling system, maintaining the capability of cooling the simulated reactor core... The results showed the rate of depressurization and coolant loss from the plant are in close agreement with the values predicted by computer codes which have been developed to predict the behavior of commercial nuclear plants under similar accident conditions..." (US NRC News Releases, March 16, 1976).

When the critics start wailing as predicted above, I will not add a footnote to this footnote.

the system would stabilize and melt no further and would be completely safe until such time as salvage operations might begin."[1]

Thus it is not the melted fuel that would be dangerous in such a disaster. The danger comes from the gaseous and volatile radioactive materials that would be released after the fuel had melted through the pressure vessel. How could this cause deaths among the public?

In most cases, it wouldn't. It is the purpose of the massive containment building — made of four feet thick and heavily reinforced concrete — to contain these gases and volatile particles within its volume and prevent a radioactive release. It is one of the formidable successive obstacles in the "defense in depth" against disaster, a defense unknown to dams, gas tanks, oil tankers and a hundred other possible causes of accidents where only a single "defense" need be punctured to result in disaster.

Could it happen that the containment building fails to contain the radioactivity within its walls? Such an event is very improbable, but it cannot be declared impossible. Since nobody knows for sure, one must postulate further possibilities, for example, a blowhole might be formed through the soil, releasing steam with radioactive material (the building itself can withstand crashing aircraft, and probably even high explosives).

After this entire chain of unlikely events, do we reach disaster at last?

No. Ordinarily, the radioactivity would be released into the atmosphere, violating NRC standards, but otherwise being dissipated into the atmosphere without significant harm. A new and independent event must arise to keep the volatile radioactive particles concentrated: a temperature inversion in the atmosphere above the plant on the day the disaster strikes. It would have to be one of the type triggering pollution alerts in cities because pollutants will not dissipate.

Only then will there be massive loss of life?

No, there still won't be any. A further independent event must occur to lead to disaster: A wind, strong enough to move the suspended particles, yet not so strong as to dissolve the inversion, must blow, and it must blow in the direction of a nearby, large and densely populated area — and not many reactors are located or planned close to populated areas; the number of people living within a 25 mile radius of current (56) and planned (44) reactor sites is 15 million, or a little over 7% of the population.

IN THE following the figures are based on the Reactor Safety Study (Wash-1400), final version of October 1975, generally known as the "Rasmussen Report," since it was directed by M.I.T. Professor Norman C. Rasmussen.

How good is the Rasmussen Report?

It involved 60 investigators, various consultants, a total of 70 man-years of effort, and about $4,000,000. Though sponsored by the AEC, the scientists and engineers working on the study came from a variety of organizations, including the AEC, private laboratories, and universities. Large digital computers processed vast data banks of information; for example, 140,000 possible combinations of radioactive release magnitude, weather type, and populations exposed were evaluated to calculate the health effects and their probabilities in a nuclear accident. The two basic techniques ("fault trees" and "event trees") used by the group have proved their worth in assessing system reliabilities in NASA and the Department of Defense; they have also been used for decades in Great Britain, where the predictions of system reliabilities were found to be close to the observed values — if anything, the techniques tend to overestimate the dangers of failure.

A draft report was published in 1974, with a whole year for critics to suggest changes. There were indeed such critics. When definite points were criticized, and material supporting the criticism was presented, the draft version was amended to take such criticism into account. For example, a study group by the American Physical Society held that Rasmussen had underestimated the number of delayed deaths, and the number of injuries due to damage to the thyroid gland. The final version did increase the corresponding number accordingly (though the risk remained minute). Similar revisions were made on the basis of criticism by the Environmental Protection Agency and some other organizations, but the amended values did not substantially change the picture given in the draft report.

The conclusions of the Rasmussen draft report, interestingly enough, were never seriously challenged by the Naderite type of critics. They had plenty of criticism, of course, but all of it abstract, vague, or just plain ridiculous. The study, they charged, did not take into account terrorism — which in itself is true; but then, not only was such a study beyond the mandate of the group, but the group had considered such combinations of circumstances as no terrorist or saboteur could ever "achieve." Ralph Nader charged that the Titanic

PALO ALTO TIMES, Friday, Nov. 22, 1974

Biologist Ehrlich assails nuclear power plants

main unresolved, Ehrlich said.

He declared that the AEC's recently published Rasmussen report on nuclear safety, officially labeled WASH-1400, "should have been called WHITEWASH 1400." The report contains serious technical flaws, he said, and its estimates of the likelihood of accidents are based on the assumption that sabotage will not occur.

Ehrlich stated "diversion of nuclear materials for radiological terrorism or the construction of clandestine atomic bombs is probably the most intractable problem associated with the nuclear power boondoggle.

"Plutonium, one of the most dangerous substances known to man, will be produced in prodigious amounts as the number of atomic power plants increases.

A "scientist" discusses 70 man-years of research.

had been claimed unsinkable, but sank nevertheless — thus demolishing a straw man of his own making, for no responsible scientist, least of all the Rasmussen group, has ever claimed that nuclear power was 100% safe.

The Union of Concerned Scientists bemoaned the fact that they did not have the Rasmussen group's $4 million to refute their report. David Dinsmore Comey, a grotesque figure even among nuclear-power baiters, came up with a number flatly contradicting the Rasmussen study by a factor of 3,000. "Comey's estimate," writes Comey, "of the probability per reactor year of a major light-water reactor accident is one in a thousand." [2] And with that, seventy man-years of effort are dismissed. Why one in a thousand? Only David Dinsmore Comey knows.

NOW THEN, suppose the entire chain of the independent and wildly improbable events described on pp.63-69 were to occur, what would be the consequences? How many people could be killed?

As in the case of all other accidents, there is no single number that will answer the question fully; the mean or expected number, for example, is so small (two fatalities) that I might be suspected of playing the game of the truth but not the whole truth. The same suspicion might arise if I selected the most probable number of fatalities in a core-melt accident (none).

The only way to answer the question fully is to give the fatalities together with their probabilities: For example, the probability of a nuclear accident, *once it has happened*, killing 10 or more people is less than 1%! (This means: In a large number of core melts, 1% will kill 10 or more members of the public, and 99% will kill less than 10, which includes none at all.) This is the probability once a core melt has taken place — it is *not* the unconditional probability of a core melt with 10 or more people being killed (which is 1 in 3 million per reactor year, or so small as to be laughable compared to the risks virtually everyone must take by merely living in America).

The probability of killing 100 or more people, again people in the vicinity of a nuclear plant *after it has had a core melt*, is 0.002. That means roughly the following: In a very large number of core melts (most of which would kill nobody, some would kill 2, some would kill more than 50, etc.), close to 0.002 of the total number would kill 100 or more.

Please note that we are concerned with the number of deaths assuming a major accident — a core melt — has taken place. We have not yet looked at a quite different question, namely, the probability that such a core melt will occur in the first place. In other words, we have looked at the consequences of an accident, not at the probability of that accident.

Let us now do just that: What is the probability of a core melt?

One in 20,000 per reactor year. And if one does happen, it will probably cost no lives, so that does not tell us much. A better way of guaging the danger is to look at the mathematical risk — the mean or expected number of fatalities per year — in comparison with other accidents. And to prevent nuclear power looking too good (it has no need for that), we will take (again from the Rasmussen Report) the averages not over all of the US, but only among the 15 million Americans who live within a 25 mile radius of the current (56) and planned (44) reactor sites:

Expected annual fatalities among 15 million people living within 25 miles of US reactor sites

Accident	*Fatalities*
Automobile	4,200
Accidental falls	1,500
Fires	560
Electrocution	90
Lightning	8
Reactor accidents	2

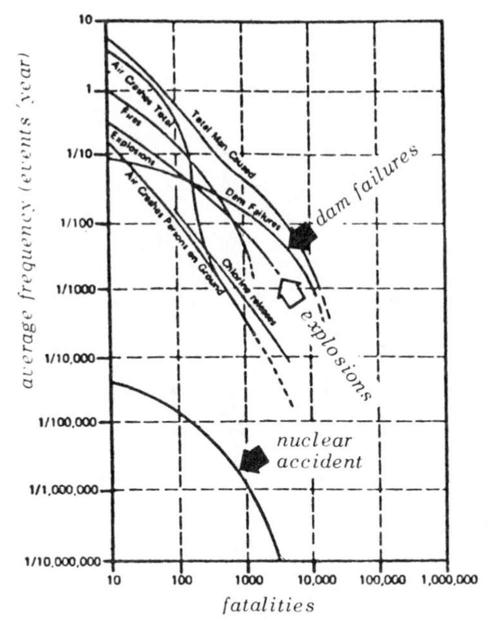

Frequency of major nuclear accidents compared with other man-caused events. From the final version of the Rasmussen Report.

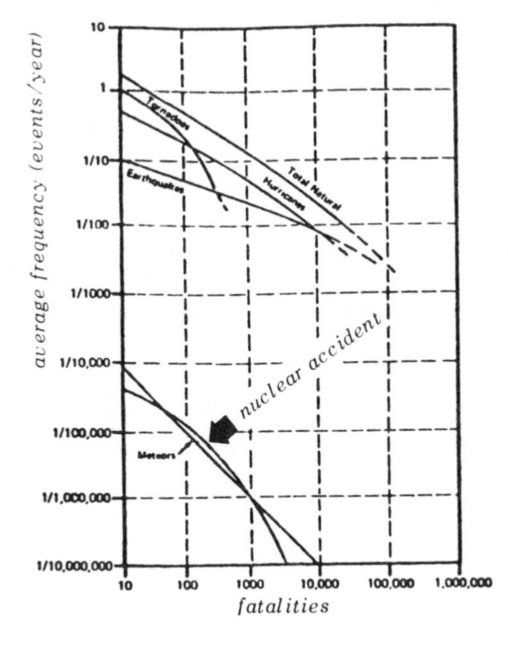

Frequency of major nuclear accidents compared with natural disasters. From the final version of the Rasmussen Report.

Here and in the following, we consider only the deaths within a short time of the accident (up to 12 months). There are no immediate deaths ("bodies piled up in the streets"), and the deaths delayed from between 10 to 40 years after the accident have already been considered (pp.59-62). The deaths considered here would be due to radiation sickness, which is fatal to some victims; those who survive it show no further symptoms of the disease.

Is it possible for a nuclear accident to kill as many as 1,000 people? It's *possible*, just as it is possible for a large enough meteor to kill 1,000 people if it were to fall into a US population center. It so happens that the probability of the two events is the same. There are only two meteors of that size known to have ever fallen on the earth, and both "missed." One lies in the Arizona desert, the other in Siberia. The probability of one of 100 nuclear plants (or a meteor) killing more than 1,000 people in a single accident is one in a million per year, which (in these two cases) is the same as once, on the average, every million years.

In short, the probability of a major accident — a core melt — is minute (about 10,000 times smaller than other accidents or natural disasters with large death tolls); and the consequences, if one does happen, are small compared to other types of accidents.

BY NOW you may be doubtful whether this is a serious book. What about the Browns Ferry fire? Didn't we almost lose Detroit? If nuclear power is so safe, why do private insurance companies refuse to insure it? And a hundred other statements will probably come to mind to anybody who ever reads a newspaper or magazine or watches television.

Yes, there was a fire at the Browns Ferry Plant in Alabama in March 1975. Yes, it was started by a candle which an inept electrician used to check whether some cables went airtight through the wall. (Since then, the NRC requires all electrical cables to have fireproof insulation.) And an electrician going about his job with a candle wasn't the only human error. Fire fighters and security were not called in until the guard sounded the alarm some 10 minutes later, and even then only after he had called the wrong number first. For several hours afterward, the plant superintendent refused to let the fire fighters use water on the fire (rather than chemicals); when the fire had been raging for 7 hours, he agreed to try water, and it was put out in 20 minutes. All of which shows, among other things,

The National Enquirer (2/17/1976)

The Horrifying Day a Blazing A-Plant Threatened 11 Million Americans

. . . People in 9 States Were Only Minutes From Death

Week Ending August 23, 1975

"Fire! Fire!" he yelled to ... electricians nearby. "Get 'em quick!" ... steel-and-concrete walls around the reactors would be blown to bits — unleashing radioactivity on nine states. Within hours hundreds of ... would be wiped out. An area would become ... fierce heat even when shut down. If water isn't kept around the core, its heat will melt down its walls — releasing awesome radioactivity.

Engineers watched, hearts in their throats, as the tremendous heat began boiling away the water ... the core

THE NATIONAL OBSERVER

The Frightful Log of a Nuclear Near-Miss

Candle Starts a Fire, Shakes Faith in Safety Systems at Power Plants

Business Week (11/17/1975).

How Browns Ferry skirted disaster

By 9:50 the water level dropped from 12 ft. to within 48 in. of the fuel rods, when a worker was finally able to repair the valves. Had he been unable to do so, the reactor core could have melted, the containment structure would probably have ruptured, and nearby Decatur, Ala., might have been decimated.

REPORT OF SPECIAL REVIEW GROUP

1.0 SUMMARY AND RECOMMENDATIONS

1.1 Introduction

On March 22, 1975, a fire was experienced at the Browns Ferry Nuclear Plant near Decatur, Alabama. The Special Review Group was established by the Executive Director for Operations of the Nuclear Regulatory Commission (NRC) soon after the fire to identify the lessons learned from this event and to make recommendations for the future in the light of these lessons. Unless further developments indicate a need to reconvene the Review Group, its task is considered complete with the publication of this report.

1.3 How Safe was the Public?

The Review Group has studied the considerable evidence now available on the Browns Ferry fire and has considered the possibility that the consequences of the event could have been more severe, even though in fact they were rather easily forestalled. It is certainly true that, in principle, degraded conditions that did not occur could have occurred. Some core cooling systems were, or became, unavailable to cool the core; others were, or became, available and some of these were used to cool the core. Much attention was drawn to the unavailability of Emergency Core Cooling Systems. While it is certainly true that the availability of these systems would have been comforting, they were not required during the Browns Ferry fire. In the absence of a loss of coolant accident, systems other than those designated as emergency core cooling systems are capable of maintaining an adequate supply of water to the core. This was indeed the case during the fire at Browns Ferry.

(published in March 1976)

how much human error a nuclear plant can take. It could, in fact, have taken a lot more.

What happened at Brown's Ferry was that some of the controls became inoperable due to the destruction of the electrical cables, but there never was a leak of the pipes, the incoming power was never lost, and the "loss of coolant" was not accidental, but intentional — after the reactor was shut down, the engineers opened the valves to condense the excess steam. One of the Emergency Core Cooling Systems had its controls rendered inoperable, but the accident never required the operation of either ECCS. One of the two units never had any trouble replacing what little water was lost, the other lost more water, but had it finally replaced from the pumps of its sister unit. At all times more than one alternative of replacing the coolant were available.*

Shortly after the Browns Ferry fire the NRC ruled that all electrical insulation in nuclear power plants must be fireproof, so that they are now safe even for people walking around with burning candles; however, the Browns Ferry fire did not breach as much as the first line of defense against a radioactive release.

The reader will find that none of this is contradicted by the nuclear critics, not even by D.D. Comey, whose reports includes all kinds of sensational details, for Mr. Comey is another of the but-not-the-whole-truth writers who doesn't contradict the truth; he just ignores it.

Comey's report fascinated a lot of know-nothings, and the media had a field day; among the journals often erroneously considered reliable, the prize must go to *Business Week*, which wrote about a worker finally being able to repair the valves (untrue). Had he been unable to do so," wrote this consistently anti-nuclear weekly, "the reactor core could have melted, the containment structure would probably have ruptured, and nearby Decatur, Ala., might have been decimated." If you read *Business Week* for its economic forecasts, its investment advice, or even its advertisements, remember that this is the journal which wrote "Decatur, Ala., might have been decimated."

And we almost lost Detroit, did we? J.G. Fuller's vile book is again based on the truth but not the whole truth; for he, too, hasn't

* Since these comments were written. they have been confirmed by the report published in March 1976 by the Special Review Group appointed by the NRC to investigate the Browns Ferry Fire (Report NUREG-0050, obtainable from National Technical Information Service, Springfield, Va.; the summary and recommendations are also given in NRC News Releases, 2 March 1976).

WE ALMOST LOST DETROIT

by John G. Fuller

This is the documented, true account of what happened on the afternoon of October 5, 1966, when the control panel inside the Enrico Fermi atomic reactor near Detroit, Michigan, suddenly registered high radiation levels, a sign of critical danger. The alarm sounded, the reactor's containment shell was automatically sealed off, and safety devices were activated. But no one knew whether the controls would hold, or whether they were facing a runaway

atomic meltdown. It was entirely possible that the reactor might explode and breach the containment building, thereby releasing enough radioactive material to destroy thousands of square miles of surrounding land. Critical questions had to be faced: Detroit was only thirty miles away. How quickly could two million people be evacuated? In which direction should they travel? What about panic? Loot-

JOHN G. FULLER is the author of the best-sellers *Incident at Exeter*, *The Interrupted Journey*, and *The Day of St. Anthony's Fire*, as well as the highly acclaimed *200,000,000 Guinea Pigs* and *Arigo: Surgeon of the Rusty Knife*. His latest book, *Fever!*, won honorable

READER'S DIGEST PRESS

$8.95

NY Times Book Review
November 30, 1975

Electricity's possible cost

We Almost Lost Detroit

By John G. Fuller.
272 pp. New York:
Reader's Digest Press. $8.95.

By MARY ELLEN GALE

"We Almost Lost Detroit" is the sort of book that could back its way into history. Its author, John G. Fuller,

When things went awry at the Enrico Fermi reactor near Detroit, four million people went about their business in happy ignorance, while the technicians gingerly tinkered with the renegade's invisible interior They knew what the public did not—a mistake could trigger a nuclear explosion.

WE ALMOST LOST DETROIT

by John G. Fuller

If this is propaganda secretly sponsored by the oil cartel to squelch the development of competitive commercial nuclear power, then we have fallen for it. It's pretty hard to remain blasé after reading this gripping account of the catastrophe that occured at the Fermi nuclear generating plant near Detroit in 1966. Its designers said it could never happen. Yet, like the Titanic that simply could not sink, the Fermi reactor ran out of control into a critical "melt down." When reactor fuel melts, no one knows what will happen next. It may drop straight through the bottom of the container and head for China, it may explode (a-la Hiroshima), or it may cool and re-solidify. Fortunately, that's what it did at Fermi number One. But re-solidified enriched uranium or plutonium in an unknown mass or shape is extremely unstable, and the slightest jar can trigger off a second dissaster far worse than the original melt-down. Which is why it took *seven months* to locate and remove *(very carefully)* the melted fuel. That's right. For seven months, the reactor was in real danger of blowing, and the resulting spreadof deadly radio-active dust and gases could have (WOULD HAVE) killed hundreds of thousands of human beings. And the people of Detroit didn't even suspect their lives were in danger.

This is an incredible story, not only about the attempts of the Atomic Energy Commission to cover up accidents like this (and there are plenty of others described in the book), but of its plans to multiply these installations all over the country. We already have fifty of them (half of those were closed down last year because of accidental radiation leaks), and more are under construction right now.

Are you still there?

We Almost Lost Detroit is an exercise in "truth but not the whole truth" journalism. Predictably, the reviews speak of nuclear explosions, reactor fuel melting and dropping straight through the "container," heading for China or "exploding a-la-Hiroshima."

raped a woman for more than a week, at least not in broad daylight. Some basic facts that Fuller never contradicts, but only very conveniently omits, have been given on p. 50, but what are the chances that the broad public will ever know them?

And yet the facts given on p. 50 are merely those that would be valid for *any* nuclear reactor; a lot of other things would have had to occur in an improbable combination before a Detroit fly got hurt.

But at the time of the incident, Fermi I could not have hurt a Detroit fly at all. The reader now knows that the danger of a meltdown does not come from the chain reaction that releases the bulk of the energy in a power plant, it comes from the fission products (p. 64). And Fermi I in October 1966, as pointed out by Prof. W. Meyer of the University of Missouri, had not been in operation long enough to have sufficient fission products to undergo a meltdown, after it was shut down, *under any circumstances.*

That means that the reviewers whose intellects are put on display on the preceding page not only write unmitigated stupidities about nuclear explosions and radioactive releases, but had they been let loose in Fermi I at the time with hacksaws, blowtorches and power drills, they could not have caused a meltdown even if they were technically literate.

Fuller's book was put out by the *Reader's Digest* Press with an advertising budget of $30,000. And it does its job. The title alone will scare many people into opposition to nuclear power. And the *New York Times Book Review* of November 30, 1975, plugged the book with a review claiming that it "is a sobering and necessary reminder that democracy has yet to control technology." The reviewer, a staff counsel of the American Civil Liberties Union, reveals a sample of her expertise in the statement "They knew what the public did not — a mistake could trigger a nuclear explosion."

As for the nuclear exclusion clause, the reason is simple: Every resident of the US is already insured against damage by nuclear plants, and with a no-fault insurance policy at that, by the Price-Anderson Act. The statement that private insurance companies will not insure nuclear plants is a pure and simple falsehood. It is two pools of private insurance companies that carry the insurance for a liability of the first $120 million per incident. The US government collects additional fees from utilities to cover *excess* liability up to $560 million. So far only the private insurance companies have paid ($400,000 in 26 minor claims not directly related to reactor incidents). The government has so far made a profit of $8 million in

Two can play the "truth but not the whole truth" game. This one is dedicated to the publishers of *We Almost Lost Detroit*, compliments of the author. For an explanation, turn the book and read the lines below.

(Virtually all foods are radioactive to some extent, particularly those containing the potassium isotope K^{40} — meat and milk, for example.)

Critical Mass (Nov. 1975) campaigned (unsuccessfully) to prevent extension of the Price-Anderson Act. Premia for nuclear liability insurance are paid by utilities to private insurance pools and to the federal government. These premia are being constantly lowered because of the nuclear plants' good record, and in any case, cost the taxpayer (*qua* taxpayer) nothing; they are ultimately paid for by the electricity consumer via the rates paid to his utility.

collected fees, and it is not likely to pay anything, as damage beyond $120 million is improbable (the Rasmussen study puts the most likely property damage to the public in a core melt at $1 million). Moreover, the 1975 extension of the Price-Anderson Act legislates a schedule of further increasing the share of private insurance, with the government eventually getting out of the nuclear insurance business altogether. Even if insurability were a good measure of safety (which it isn't), the nuclear critics would not have a point, for their charges are simply not true.

One might also add that in the case of *all* other disasters, the no-fault insurance liability limit is not $560 million; there isn't any no-fault insurance at all.

There are, of course, a thousand more 10-second statements by the nuclear critics that it takes a half-hour lecture each to refute fully, but it is hoped that these three samples will have persuaded the reader that the consequences and probabilities of a nuclear accident described above are credible in spite of the wild claims of the critics. In fact, the discrepancies are rarely contradictions; they are most often due to facts which the critics do not dispute, but on which they conveniently keep silent.

But this is not a book attempting to refute the critics' 10-second assertions; it is a book on the hazards of *not* going nuclear, so let us move on to the major accidents and disasters due to fossil fuels and hydropower.

BEFORE we go into the possibilities — and indeed, the actual occurrences — of major accidents associated with fossil fuels, let us first clear the ground of some misconceptions that confuse the issue. The death toll in automobile accidents, airline crashes, railroad disasters, etc. (usually) has nothing to do with the generation of electrical power, and the table on p. 70 had no other purpose than to let the reader guage the numerical value of the risk associated with a nuclear disaster. There is no choice of either being struck by lightning or perishing in a nuclear disaster; one can take measures to minimize the dangers, but the two types of measure (say, installing a lightning conductor and voting for a nuclear shutdown initiative) do not involve mutually dependent decisions, let alone mutually exclusive ones.

The anti-nuclear crusaders are therefore right, in my opinion, when they say that it is unfair to compare the risks in driving an automobile to the risks of a nuclear disaster. Unfortunately, their thinking is so muddle-headed that they are right for the wrong reason. "Exposure to the risk of auto accidents," says Sierra Club Executive Director McCloskey,[3] "is a personal one accepted by those who choose to travel by car; it is not forced unknowingly on an entire population regardless of their choice." That, of course, is a fallacy. If Mr. McCloskey's wife has a baby, how much choice does she really have, in the United States, whether to be driven several miles to the nearest hospital or whether to walk there? The real reason why the comparison is unfair has escaped Mr. McCloskey. It is that whether we do or do not go nuclear will make no significant difference to the death toll on the highways.

But it will make an appreciable difference to the death toll in coal mines and in major accidents due to the explosion of oil and natural gas tanks, oil refineries and oil tankers, for the simple reason that a considerable fraction of these fossil fuels is used as fuel in *power plants*. If x percent of the US electrical generating capacity goes nuclear, roughly x percent of these fossils burned to generate power will not be needed, and the death toll in major accidents associated with the production and storage of these fuels will decrease, on an average, by the same percentage.

In the following, we shall therefore not make unfair comparisons. True, the risk in flying with commercial airlines is not only high, but partly unnecessary. It is high because per passenger-hour it is even greater than driving. (The airlines conveniently give the risk not in passenger-hours, but passenger-miles; by that measure, the safest mode of transportation is a NASA flight to the moon.) And it is unnecessarily high because to this day airliners do not have airborne radars to warn them of threatening collisions, and because the political clout of the environmentalists forces pilots to throttle back their engines dangerously early after take-off. But airlines are none of our business here, for we are interested only in the hazards associated with the alternatives to nuclear power, of which air travel is not one.

AND YET we must come back to the airlines one last time, for air crashes do have something in common with mine disasters and oil tank explosions that is relevant to our story. And that is the indifference — one might almost be tempted to say callousness — with which news of such disasters is received by the public. No year goes by without a major air disaster. 10 dead in a private aircraft crashed in the Rockies; 83 dead in a crash in Florida; 102 dead in a crash in New York; 132 dead in a crash in Lebanon; it has become too common to startle people. Even in the United States major crashes have become so common that a year in which none happens, such as 1974, makes the news by its rarity.

The crashes make the news, too, of course. The TV screen shows the debris scattered in a forest, rescue crews cutting the fuselage with welding torches, and covered bodies being carried away on stretchers. But the TV viewer has seen all this too often before, and the horror of it is forgotten with the next swig from his beer can.

So, too, forgotten are the 400 coal miners killed in a mine disaster in India in December 1975. India, of course, is far away; but how about West Virginia? Only 8 years ago, 78 miners died in the Mannington Disaster of 1968; "of 1968" to distinguish it from the Mannington Disaster of 1907, when 359 people died. Thirty-eight miners lost their lives in a coal mine explosion near Hayden, Ky., in 1971; and it won't be long before another mine disaster costing tens of lives comes along with the inexorable laws of probability, here in the United States.*

* It came along before the manuscript of this book went to press. Twenty-six

After which our TV viewer will take another swig from his beer can.

Is anything of the kind thinkable in the case of nuclear power? The Browns Ferry fire, in which there was not a single casualty, is alive and well in the twisted news columns after more than a year; the two out of 100 fuel rods of Fermi I, which caused the safety equipment to work exactly as intended, are transformed into the We-almost-lost-Detroit baloney a decade after the event. The two hairline cracks in the ECCS pipes in the Dresden I plant in Illinois, too small to leak moisture, were blown up into a major event by an irresponsible and ignorant press.

For hundreds of reactor-years, there has not been a single reactor-related fatality in the generation of commercial power anywhere in the United States. But by the same inexorable laws of probability, one day such a fatality will take place; perhaps it will have taken place by the time this book is published. It is certain that this time the casualty or casualties will rate more than another swig from the beer can. In such an event, one can only hope that the American people will not be driven into a blind panic by a small group of vociferous witch hunters, but that they will keep their sense of proportions.

A sense of proportions is what this book is about.

WHEN one considers the major accidents associated with the fuel cycle of fossil-burning power plants, one difference with respect to nuclear accidents stares one in the face: The fossil accident statistics have not come out of computer simulations of hypothetical disasters, and they have not come from probabilistic calculations. They have come from the cold records of the coroners.

There is, as yet, only one type of nuclear accident: hypothetical. There are two types of non-nuclear accidents: hypothetical and real. We begin with the real accidents.

Coal mining, as everyone knows, is a dangerous occupation; but few of those who lecture us on the hazards of nuclear power have ever been down a coal mine, and fewer still realize that the miner who has escaped violent death by methane explosion, flooding or collapse of the walls faces even greater hazards from contracting occupational diseases such as Black Lung.

---

miners and rescue workers were killed by methane explosions in the No. 1 Black Mountain Mine in southeastern Kentucky in March 1976.

Since 1907, no less than 88,000 miners have died in American coal mines, and even now there are some 200 fatal accidents per year in coal mines, plus another 100 in transporting the coal to the power plants.[4] The average number of fatal coal mining accidents for the 5 years 1965-1969 was 246 per year. There were only 8 fatal uranium mining fatalities per year during the same period, but that is merely due to the fact that far less uranium is mined than coal; as far as the *rate* of accidents is concerned, the two are about equally dangerous (injuries per million man-hours are 43.5 for coal and 39.8 for uranium, and disability days per million man-hours are 8,441 for coal and 8,702 for uranium).[5]

But the question of interest here is this: What is the cost, in accidental mining deaths and injuries, of the production of a given amount of electric power? The answer depends on many factors, but it is dominated by a single aspect: the concentration of energy in a given quantity of fuel. A pound of unrefined uranium ore contains about 100 times as much energy as a pound of coal, so that about 100 times more coal must be mined to generate the same amount of electric power, and one would expect the cost in lives and injuries to be of the same order.

Accidental deaths in mining, per electric energy produced from the corresponding fuel, are about 10 times more numerous for coal than for uranium. Lave and Freeburg[5] investigated the data for 1969, when 54.3% of the mined coal was used to generate 705 million MWh of electric power, and about 3.06% of the mined uranium was used to generate 14 million MWh of nuclear power. There were 8 fatalities in uranium mining (plus an average of 1 fatality in 5 years in uranium milling, which has no equivalent in the coal cycle). The figures we are after, then, are the following:

Per billion MWh of electric power consumed, the cost in fatal mining accidents is

189 lives in coal mining for coal-fired power

18 lives in uranium mining for nuclear power

which is close to 1 : 10 in favor of nuclear power. (The figure for uranium mining is actually 17.92; however, due to weapons production, storage, non-linear consumption of uranium in the fuel rods, and other factors, the calculation of the power-destined fraction of the mined uranium ore is less accurate than for coal.)

The ratio of 10 : 1 also holds for injuries:

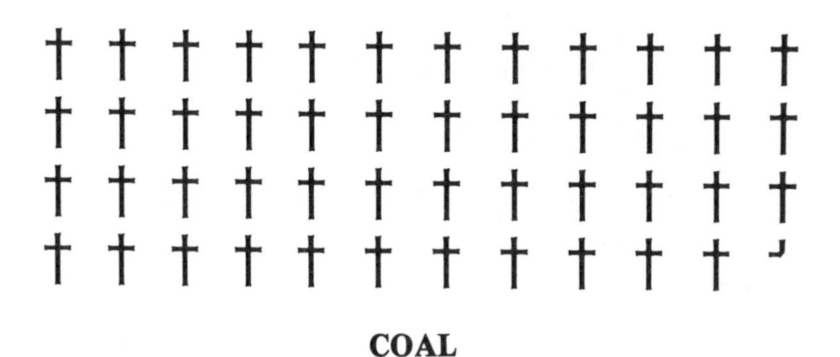

Fatal accidents in mining. For the same electric energy produced, there are close to 100 times more fatal accidents in mining coal than in mining uranium. If the energy produced is 1 billion MWh, each cross represents 4 lost lives.

Per million (not billion as above) *MWh of electric energy consumed, injuries cost*

1545 disability days among coal miners for coal-fired power
157 disability days among uranium miners for nuclear power

But these are only accidents; and miners are also subject to industrial diseases. Coal miners contract, above all, pneumoconiosis (Black Lung), and uranium miners have a higher incidence of cancer than the average citizen, since they are exposed to radiation (uranium, with its long halflife, is not harmful, but its daughters, particularly radon, are). The toll of these diseases among miners is far greater than that of accidents: There are about 4,000 deaths by Black Lung per year among coal miners, and about 20 deaths by excess incidence of cancer per year among uranium miners. However, it is very much more difficult to estimate the toll per unit power generated, and it is not even easy to estimate the ratio of deaths by occupational diseases per unit power generated for coal versus uranium. (Among the reasons is the delayed and prolonged time of a disease, whereas accidental death or injury strikes on a single day.) We shall be content with the estimate of Prof. Richard Wilson of Harvard University, who is a specialist on the epidemiology associated with fuel cycles. Wilson's estimate[6] is the following:

Per billion MWh of electrical energy consumed, there are
1,000 deaths by Black Lung among coal miners
20 deaths by excess lung cancers among uranium miners
depending on whether the power is coal-fired or nuclear.

In fairness, it should be pointed out that these two figures partly rest on debatable assumptions for the reasons pointed out above, and that they are not as accurate as the figures on accidental deaths and injuries. On the other hand, the discrepancies between the results of various investigators involve only the degree to which coal is more dangerous; no one disputes that coal is the much more dangerous of the two. (Wilson's estimate above implies a ratio of 50 : 1; the lowest estimate I have seen is that by Lave and Freeburg,[5] which is 18 : 1.)

BEFORE we go on to transportation accidents (which kill uninvolved members of the public, not only people who have chosen to become miners or railroad workers), we pause to consider the following argument:

"It is deplorable that mining is such a dangerous occupation; but every miner knows well what the risks are. In the United States, at least, nobody is forced to be a miner, and if somebody chooses to be a miner rather than a shepherd, the choice is his, the responsibility is his, and the consequences are his only."

This type of argument is debated in a hundred versions between libertarians (who believe in freedom of choice and individual responsibility) and liberals (a misnomer for people who believe that all of society is responsible for all of its individuals), and they will argue the point for hours without noticing that the entire debate is pointless because it is based on a false premise.

One can argue the point in the case of a tightrope walker or a person who chooses to go over the Niagara Falls in a barrel, for apart from the bother to a rescue crew and perhaps burial at public expense, the consequences for the rest of us are minimal. But a miner is not a tightrope walker, and the false premise is that the

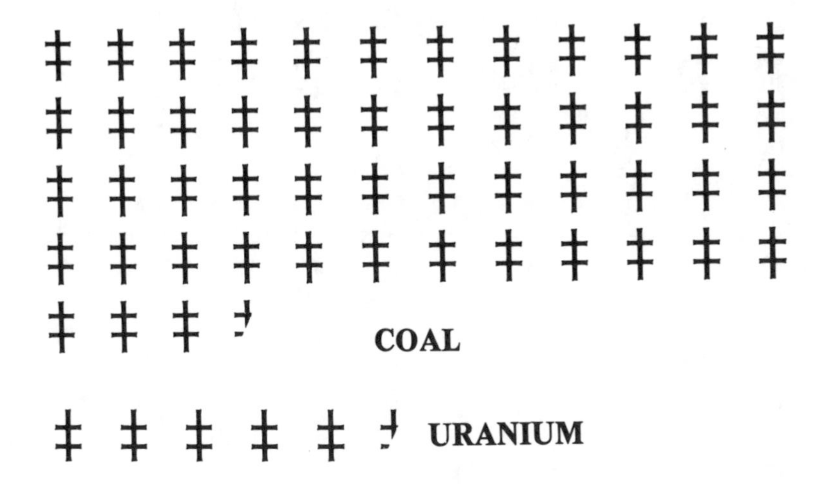

Injuries in mining. If the electrical energy produced from the corresponding fuel is 1 million MWh, each symbol ‡ represents 30 disability days.

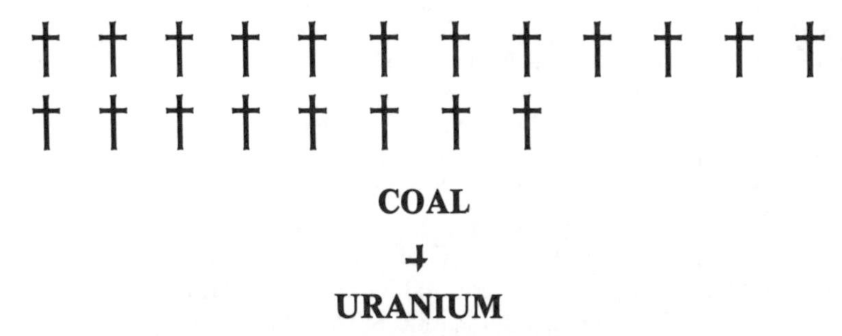

Industrial diseases: Black Lung among coal miners vs. excess lung cancers among uranium miners. If the electrical energy produced from the corresponding fuel is 1 billion MWh, each cross represents 50 lost lives.

consequences of his choice are borne by him only. They are not, by any means. Never mind the moral aspect, never mind the warring scholars of philosophy, just consider this single fact: The US government now pays close to $1 billion a year in support of black lung victims, of which there are 50,000 alive at any one time. "The US government pays" means, of course, that the taxpayer pays; it comes out of your pocket and mine. (Incidentally, though it is not the subject of this book, the government subsidies to the nuclear in-

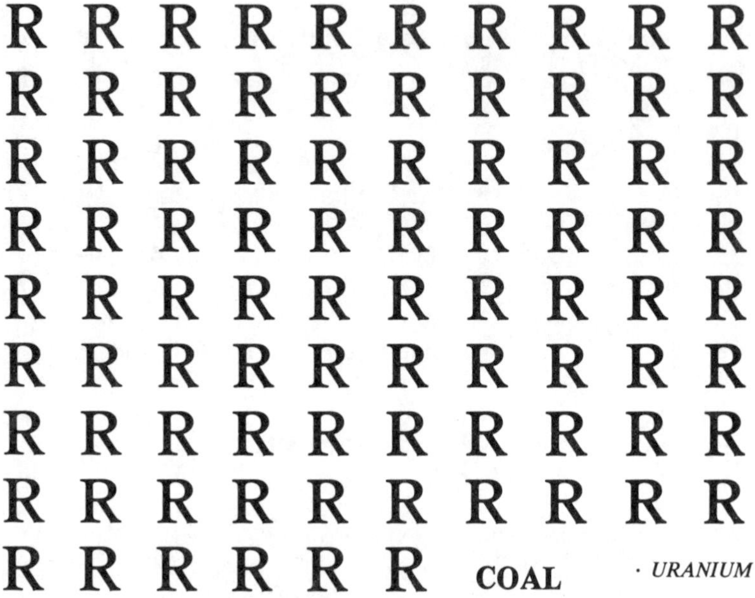

R R R R R R R R R R
R R R R R R R R R R
R R R R R R R R R R
R R R R R R R R R R
R R R R R R R R R R
R R R R R R R R R R
R R R R R R R R R R
R R R R R R R R R R
R R R R R R **COAL** · *URANIUM*

In transporting coal from mine to power plants, about 100 people (including members of the public) are killed every year. The corresponding number for uranium is unknown and probably very close to zero. The reason for the disproportion is the high concentration of energy in nuclear fuel: It takes 38,000 railcars of coal, but only 6 truckloads of nuclear fuel, to supply a 1,000 MW plant for one year. For this case, each "R" represents 500 railcars of fuel.

dustry are another myth; but not even in Nader's mythology do the alleged subsidies amount to $1 billion per year.)

If $1 billion of public money were annually paid out to tightrope walkers, the American taxpayer (one hopes) would object; but in the case of Black Lung, he does not realize that it is a disease whose toll can be cut: Every percentage point by which nuclear power takes over from coal saves approximately 20 human lives from death by Black Lung — not by curing it, but by preventing it. It saves them in addition to the lives saved from death and disablement by violent accidents. And in addition to the lives lost by causes which we are about to examine.

So for those who feel that they are not the guardian angels of coal miners, there is always the prorated amount of $1 billion a year ($80 million additional if the present nuclear capacity went back to coal).

Ralph Nader's brand of humanitarianism couldn't care less about coal miners coughing out their lungs; but if people are so hardboiled that they have no compassion for the miners (and I do not intend to moralize), shouldn't they have a little compassion for their own pocket books?

IN any case, once we start looking at transportation, the point becomes moot, because fatalities in fuel transportation involve members of the general public, even if to a lesser degree than truck drivers and railroad workers; the element of choice begins to disappear. There are 100 accidental deaths a year in transporting coal from mine to power station;[7] as for the number of deaths in transporting uranium ore from mine to mill and refinery, and from there (as "yellow cake") to enrichment plant, I have, frankly, been unable to ascertain it, and I suspect that it is so small as to elude the record keepers. But when transportation of ready fuel to the power plants is considered, the figures are truly stunning, and they also make it obvious where the 100 deaths in transporting coal to power plants come from. To generate 1,000 MW-years of electricity (this would in practice correspond to the output of two 1,000 MW units working with a plant factor of 50% for one year), the amount of fuel that must be transported to the plant is

either 6 truckloads of nuclear fuel
or 38,000 railcars of coal.[8]

I will not insult the intelligence of the reader by commenting on these figures.

THE 4,300 people who die in accidents in producing and transporting coal (not counting Black Lung victims) are a small amount compared to the far larger number who die as a direct consequence of air pollution by the combustion products of coal. There is, however, one thing one can say in favor of the safety of coal: Except when pulverized and suspended in air, coal will not explode, and apart from a few fires, it will not cause violent death to large numbers of the public.

The same cannot be said for gas or oil, where the situation is the opposite: Few lives are lost in producing oil and gas, but their transportation and storage are extremely dangerous.

Per unit electricity produced by burning oil, the number of deaths and injuries in producing and refining oil is quite small; it is even somewhat smaller than the corresponding figures for uranium mining and milling. For example, Lave and Freeburg[5] found 135 disability days per billion MWh of oil-generated electricity (compared to 157 in uranium mining). But the dangers in storage are another matter.

An oil-fired power plant of 1,000 MW capacity burns 40,000 barrels of oil a day. It usually keeps a six weeks' supply on hand, which works out to 2 million barrels of oil. What would happen to the public if such a large amount of oil caught fire is, in one respect, remarkably similar to the case of a nuclear disaster, in that the consequences would depend on the meteorological situation. As in the case of a nuclear disaster, the consequences are worst when the smoke is kept close to the ground by a temperature inversion and a light wind blows it toward a large population center. In that case thousands could die by asphyxiation and by induced or exacerbated lung and bronchial diseases (not counting possible lung cancers).

Such a nightmare was close to coming true on January 6, 1973, when an oil fire started in Bayonne, New Jersey, as a result of the collision of two ships, with oil storage tanks catching fire on shore. The black smoke (see photo on opposite page) was far denser than the notorious air pollution of London in December 1952, which resulted in 3,900 excess deaths there. Fortunately, the wind blew away from Manhattan and was turbulent enough to disperse the smoke, preventing large-scale loss of life. But nobody wrote a book called *We Almost Lost New York.*

If Ralph Nader and his fellow-humanitarians had the slightest interest in safety or human lives, they would have pounced on this near-disaster. They are, after all, so touchingly concerned about this type of catastrophe when it is nuclear and has a probability of once in a million years.

But it didn't take a million years before New York had another close shave of this type. It only took three years, almost to the day: On January 3, 1976, a 90,000-barrel oil storage complex in South Brooklyn caught fire and exploded; but this time it was worse. The fire could not be brought under control, and the next day, a second explosion in the burning tank belched forth a gigantic orange fire ball, igniting a second tank and rupturing a third. The three tanks, though not full, contained a total of 3.1 million gallons of oil. The fire raged several days and was not brought under control until January 7.

We almost Lost New York. Oil fire in Bayonne, N.J., in the New York metropolitan area, on January 6, 1973. The black smoke was far denser than the air pollution in London in December 1952, which resulted in 3,900 excess deaths. As in the case of radioactive clouds, the danger is dependent on the weather; this time, Manhattan escaped disaster by the absence of an inversion and unfavorable wind direction.

Several persons, mostly firemen fighting the blaze, were killed; but here we are concerned with massive loss of life to the public. Once again, New York was saved by favorable weather: The gigantic plume of thick, black oil smoke rose high into the sky (see photo on p. 91). Had there been an inversion to trap it, and a slight wind blowing it in any direction except out to the ocean, several thousand people would have died, as they did in London in 1952 under much less severe air pollution. Once again, nobody wrote a book called *We Almost Lost New York*. Once again, few people noticed the similarity with a nuclear disaster.

Typically, *Time* did not consider the event newsworthy at all; but two months later,[9] it again harped on the Browns Ferry fire and brought a series of diagrams labeled "Three Steps to Meltdown:" 1) loss of coolant, 2) failure of the backup system (ECCS), 3) meltdown. The third diagram depicts the containment building ruptured and radioactivity escaping, with no explanation why it should. In reality, rupture of the containment building is a fourth and independent step, but not only did *Time*'s editors get the four steps garbled, they forgot about the weather, which would have to be such as to get the radioactive cloud into a population center. More significantly, they never noticed that two months earlier the oil fire was the equivalent of having arrived at the fifth and last step, and in the middle of a densely populated area, too, namely, in Brooklyn, a few miles from their editorial offices. (I do not believe *Time*'s editors did this on purpose; which is to say, I believe they are grossly incompetent.)

But similar as an oil fire and a radioactive release may be in causing massive deaths to the public, the differences are even more striking in underlining the hazards of the *former*. There is, first of all, no defense in depth for an oil storage complex; there is no equivalent of an ECCS, a steel pressure vessel, or a massive ferroconcrete containment building. An oil storage complex only needs something to ignite it, and that's it.

Second, the probability of an oil fire is not just "greater" than that of a nuclear accident *that kills the same number of people* (whether 3 or 300), but it is greater by a factor in the tens of thousands.

Third, there just isn't a nuclear plant smack in the middle of New York City; the law forbids nuclear plants in or near large population centers. Grand Central Station in Manhattan would break the law if it were a nuclear plant, for even the radioactivity of its granite blocks exceeds NRC standards. Note, however, that I am neither advocating the removal of oil storage complexes (let alone Grand Central Station) from New York City, nor am I opposing it; I am just comparing things with the same yardstick.

But suppose that by some miracle Ralph Nader suddenly got interested in protecting consumers or advocating safety, and that he would use his anti-nuclear lobby for getting oil storage complexes out and away from New York City and other urban areas; would the risks of oil fires drop below those of a nuclear accident?

No, they wouldn't. The risks from oil fires at power plant storage facilities only — disregarding the tankers and transit storage necessary to get the oil there — are still higher than those due to a nuclear accident; to put it more accurately, the consequences of such accidents are worse than for a nuclear disaster of equal probability. Using American Petroleum Institute fire statistics, Starr and co-investigators[10] estimate the probability of an oil fire with 10 or more deaths to the public at 1 in 10,000, which is twice as high as the probability of a correspondingly severe nuclear accident, even if there

SMOKE BILLOWS OVER BROOKLYN — A plume of smoke billows into the sky over the burning Patchogue Oil Terminal Corp. in South Brooklyn after a new explosion Monday rocked the oil tank storage area. Air view is looking toward the East. Waters are those of the Gowanus Canal, an inlet of Upper New York Bay and a major route for fuel and industrial supplies

(AP Wirephoto)

We Almost Lost New York a second time. Near-disaster struck closer and more ominously three years later, this time in Brooklyn. New York escaped once more due to favorable weather conditions, the same way a city might escape disaster in a major nuclear accident.

According to PIRG attorney Louis Sirico, "No other nuclear power plant is in so densely populated an area as Indian Point. There are more than 50,000 people within a five-mile radius of the facilties, more than

From an issue of *Critical Mass.* Apart from the beautiful picture, note the text underneath. PIRG (Public Interest Research Group) is one of the Naderite organizations claiming to be concerned about public safety. It is remarkably worried about the Indian Point plant with 50,000 people in a five mile radius, but was remarkably indifferent to the January 1976 Brooklyn oil fire with several million people in an equally large radius.

were 100 nuclear power plants in the country. (As of April 1976, there were 60 licensed reactors, not all of them on line yet.)

This, however, is only the oil stored at power stations. What of the oil stored in far larger storage facilities in ports, near refineries and other places? The oil stored at power plants, after all, must have gone through some of these places first.

There is, for example, a place on the East Coast — well known to researchers on the subject, but I am reluctant to reveal it here — where 151 million gallons of oil are stored literally on top of a town with a population of 37,000 people. What if the complex caught fire and the tanks exploded? (Whether a saboteur could do as much damage with a home-made plutonium oxide bomb is highly dubious, but it is certain that he could not do it as quickly and safely for himself.)

It should also be remembered that oil tank explosions and fires are not, like nuclear accidents, hypothetical; they take place all the time. But if you do want to speculate about what *might* happen, consider the energy carried by an oil tanker. A fully laden 200,000 ton oil tanker carries the energy of a two-megaton hydrogen bomb. There are about 60 of these tankers now in service, and others, with 500,000-ton capacities are under construction.[11] The US does not yet have a deep-water port that could receive one of these supertankers, nor would the energy be released in a small fraction of a second, as is the case with a nuclear bomb. So if a regular oil tanker should explode in Seattle or Baltimore, the consequences would not be as bad as a two-megaton explosion of TNT, and we can all sleep soundly again: The energy released would be no more than a couple of nuclear bombs like the one dropped on Hiroshima...

NATURAL gas is even worse for major disasters, especially in the form now most often used for storage, i.e., as liquid natural gas (LNG). Ships containing 1 billion cubic feet of liquefied methane regularly dock in heavily populated areas — for example, in Everett, Mass., 1½ miles from downtown Boston.[12] To see what would happen if one of these ships exploded, or at least started a big fire, one needs to do little more than strike a match. Ralph Nader's heart does not bleed for the potential victims.

Potential victims? The victims of major LNG disasters are not potential, they lie dead in the ground, unbewept by the anti-nuclear hypocrites. In October 1944, LNG tanks exploded in Cleveland, Ohio, killing 133 people, some of them after the LNG had entered sewers and caused fires. Could an accident of this sort happen today? Yes, and on a larger scale: The volume of LNG involved in Cleveland was only 50 million cubic feet; today's LNG tanks store 20 times more.

On February 10, 1973, an *empty* LNG tank exploded and its lining burned.[12] Fortunately, the tank was empty, so that only 33 people were killed...

In minor accidents with gas, about 100 people a year lose their lives (10 lives per billion MWh of consumed energy, not necessarily electrical). Once again, these are not hypothetical deaths calculated via fault trees and computer simulations. They are corpses counted by American coroners.

Hypothetical speculations of what LNG could do via tanker explosions in ports, or via storage tanks in cities, or via explosion of the underground network beneath, say, Manhattan island, could make the worst hypothetical nuclear accident look like a picnic, and we will leave this type of fantasy to the script writers of Hollywood's next disaster film. However, Prof. Richard Wilson of Harvard University has made an interesting comparison on the money spent to save a human life from LNG tank explosions versus the money spent to save a human life from the radioactive emissions of nuclear power plants.[13]

When, in 1973, the maximum permissible radiation dose at the property line of a nuclear power plant was reduced from 170 mrems/year to 10 mrems/year, the effect was to reduce the incidence of cancer from 4 to 1 per year (out of a total of 300,000 cancers in the US). The cost of this step worked out to $800,000,000 per saved life.

On the other hand, there are now 75 LNG tanks located in US cities. The cost of moving these tanks out of the cities (calculated in the same way as for the example above) would amount to only $1,000 per saved life; but this cost has *not* been paid, and the LNG tanks remain in the cities.

Now who is it (and here we are no longer quoting Prof. Wilson) that decides to pay $800,000,000 for saving a human life from one danger, but refuses to pay $1,000 to save it from another?

In legalistic theory, it is the agencies of the US government, by the power delegated to them by the American people. But in reality, of course, we know better. We know that the vast mass of the American people does not know about millirems or LNG tanks, and cares even less. And we know that technical decisions of this type are made, whether by politicians or bureaucrats, under the prodding of pressure groups and lobbies. The anti-nuclear movement has become a powerful political force; there is no comparable movement to get the LNG tanks out of town.

Why not? The answer is painfully obvious. Moving the LNG tanks will do no more than save human lives. But attacking (ridiculously low) radiation levels is a scare tactic that can be used as a crusading horse against the big corporations and the "establishment," and hence for garnering political power.

FOSSIL fuels, then, present a far larger risk of accident, both by their greater consequences and by their greater probabilities, than does nuclear power. And fossil fuels are the only alternative that can completely *replace* nuclear power; in fact, with oil and gas running out, only coal can replace it completely.

Other sources, such as solar or hydropower can supplement the basic energy sources, but they cannot replace them. There are not enough sites for hydropower left in the US to make a decisive difference, and solar power, as we shall see in a moment, can provide at best a small fraction of the total need. Wind, geothermal and tidal power make good conversation pieces, but none of them can reasonably provide more than 1% of the US need, and we shall waste no more space on them (they, too, are more dangerous than nuclear power per energy produced, and environmentally less satisfactory).

Because of their quantitative limitations, solar and hydropower would be out of the running even if they were safer.

But they are not. Dams are no safer than other energy facilities, and they are grossly less safe than a nuclear power plant. Dams do break and kill people, with a probability about 10,000 times greater than nuclear power plants. A dam failure killing 1,000 people is estimated to occur, on the average, every 80 years; a nuclear disaster of that magnitude, according to the Rasmussen study, would take place once in a million years.

A recent study at the University of California at Los Angeles revealed that the failure of certain dams in the US could could cause tens of thousands of deaths, and one of them could cause between 125,000 and 200,000 fatalities.[16] (The Rasmussen study stops at 3,000 deaths, where probabilities are already absurdly small.)

In March 1928, the St. Francis Dam in Santa Paula, California, collapsed and killed 450; in December 1959, the Malpasset dam in France collapsed, killing 412; in February 1972, coal mine waste waters caused a makeshift dam to collapse in Buffalo Creek, West Virginia, killing 118 people.

On October 9, 1963, there was a dam disaster in Vaiont near Belluno, Italy, which was not caused by the dam actually breaking, but by a mountainside collapsing into the reservoir and flooding the valley below (just as if the dam had broken). More than 2,000 people were killed, and 50,000 were left homeless.

In the 1971 Los Angeles earthquake, a dam above the San Fernando Valley cracked and would doubtlessly have given way had the reservoir been full of water; but it so happened that due to the high demand for electricity on that early February morning, it was partially empty. Someone forgot to write a book *We Almost Lost Los Angeles.*

But then there is solar power, the good guy, the one that is not used "for only one reason — that the oil companies do not own the sun" (Ralph Nader, of course).

"Do you want a sunshine future for your children, or a radioactive one?" ask the advertisements of Environmental Action.

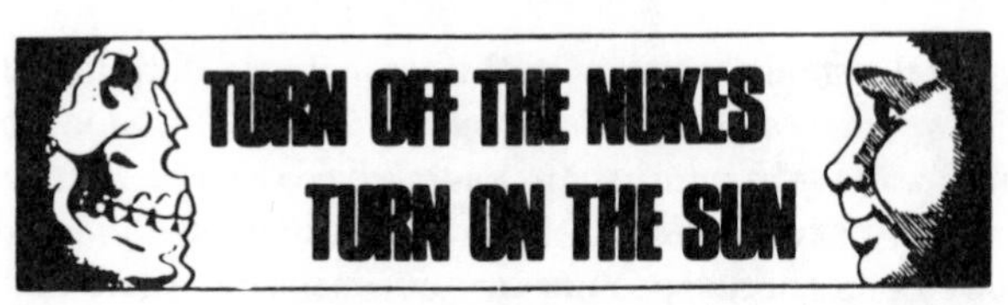

Bumper sticker sold by *Environmental Action.* Solar power is not particularly safe; see text why it can never hope to match nuclear safety.

The good people of Environmental Action will never believe this, but nuclear power is far safer than solar power, too; solar power, in fact, is not particularly safe at all.

There is a very good reason why nuclear power has had such an unparalleled safety record. The strict regulation by the AEC (now NRC) has, no doubt, helped, but it is hardly the major underlying reason. If it were that simple to legislate safety, there would not be 50,000 dead on US highways every year.

The underlying reason is quite different: No other power but nuclear has its dangers so concentrated in a very small space — the reactor core. No other fuel but nuclear has its energy so concentrated in a tiny volume. One pound of plutonium has the same energy as the Yankee Stadium full of coal. One hundred people a year are killed merely by transporting 150 million tons of coal all over the country; but virtually nobody is killed by a few truckloads of nuclear fuel.

There is no way of controlling a diluted danger such as presented by 105 million cars on the roads; but one can come close to perfect safety (though never completely achieving it) when the danger is all locked into a few cubic yards of space. It is the great concentration of the danger, and hence the comparative ease of guarding against it, that is the underlying reason for the unusually high degree of nuclear safety.

In that sense, solar power is the very opposite of nuclear power. The salient feature of solar power — one that surfaces in safety, economy, effectiveness, environmental impact, and all other aspects — is its diluteness. At the best of times, on a cloudless day with the rays hitting the collectors perpendicularly, the *influx* of solar energy is 1 kW per square meter (10.4 square feet), which means that large amounts of power need large collecting areas. For example, with 10% efficiency and 50% spacing (betwen the collectors), a 1,000 MW solar plant works out to no less than 50 square miles of collecting area (compared to a few acres for a 1,000 MW fossil-burning or nuclear plant). What that means to economy and environmental impact is another matter, but what does it mean for the danger of accidents?

First of all, 50 square miles of area per power plant unit are not easy to come by where electric power is most needed today, in the northeastern United States. They are available in, say, the Arizona desert, but it is uneconomical to transmit electric power over very large distances; most probably the solar plant would produce hydrogen (by electrolysis of water) for shipment, and we are back to square one. Transportation in large quantities (except by pipeline) means accidents in large quantities; hydrogen is flammable, and mixed with oxygen, i.e., with air, it is explosive. Like methane, it is liquefied for storage and transportation, and we are simply back to the dangers of LNG and natural gas in general.

But where the diluteness of solar power really touches on safety questions is the 50 square miles of collecting area itself. Fifty square miles will be covered with large structures, which have to be maintained. They have to be kept clean, for one thing — clean of dust in Arizona, and clean of snow in Nevada. And there are going to be plenty of accidents on that 50 square mile area of large structures, especially accidents by the no. 2 killer in the US: accidental falls.

Everybody knows the no. 1 killer among accidents, automobile accidents, with a toll of about 50,000 deaths per year. But the no. 2 killer is not only deadly, it is also unknown to most people. Most people, when asked, guess at the small fry: fires (6,500), explosions (500), electrocution (1,000), firearms (2,600), drowning (7,000), air travel (1,700), tornadoes and hurricanes (200), and so on. But none of these amount to very much unless they are all lumped together as "other accidents." The no. 2 killer kills about 16,500 Americans a year (it used to be 20,000 a year until recently, when the Bureau of the Census modified its classification of accidental falls), and when, for solar collecting structures, we add 1,300 deaths by blows from falling objects, we have a total of nearly 18,000, or more than twice the next biggest group (accidental drowning). [15]

Not that this makes solar power particularly dangerous. The danger doesn't even constitute the main objection to solar power. But there is no way one can watch over 50 square miles the way one can watch over a single reactor vessel inside a ferro-concrete containment building. Even if a single accident can be much worse for nuclear than for solar power, the risk (probability times fatalities) will still be smaller for nuclear.

That is the way things look for large-scale solar energy conversion. For small-scale conversion by homeowners, the situation changes significantly — for the worse. Suppose solar-electric conversion came down from its present investment cost of $1,500 per kW so that people could afford it, and suppose they were willing to assume the investment costs (rather than pay about 4 cents per kWh to the utility, with everything taken care of). Then it will be laymen, not professionals, climbing on the roofs to clean off the snow after every storm, and they will climb onto roofs, not collector structures built for that purpose in large-scale conversion.

But above all, they will be risking their lives, not for 1,000 MW of capacity, but for a crummy 5 or 10 kW. One cannot produce a billion MWh by piddling around with 5 kW at a time, but one can use the same unit for the *rate*, deaths per billion MWh, as we used for coal and uranium mining and other parts of the energy cycle. What that number would be exactly, nobody knows; what is certain is that it would be of carnage proportions.

And we haven't even considered accidents with storage facilities — sulfuric acid in the batteries, the explosive hydrogen-oxygen mixture arising in charging them, etc., all in the basement, laundry room or otherwise exposed for children to play with.

Solar power may come down in price, and all kinds of technology may be invented for it. But one thing can never change, and that is the incoming 1 kW per square meter. It is this diluteness of solar power which is the ultimate cause of its lack of safety.

Please do not misunderstand me. I am not saying that solar energy is a bad thing, and I am not even saying that safety should be the only consideration in the choice of an energy source. What I am saying is that solar energy doesn't have a chance of even coming close to the safety of nuclear power.

4

Waste Disposal

Reactor safety was a good issue to jolt the public, but compared to the issues of radioactive waste. . . it is like estimating how many angels can dance on a pin. The decision to obtain two percent of our energy in barter for the human gene pool is morally indefensible and a national abdication of morality.

Lorna Salzman, Mid-Atlantic Representative of the Friends of the Earth.

Waste disposal, so often touted as a bogey by the anti-nuclear crusaders, is in fact one of the prime reasons why nuclear power is very much preferable to coal-fired power. If all of the US power capacity were nuclear, the total amount of wastes per person per year would amount to one aspirin tablet, and that can easily be disposed of deep in the earth, where it came from in the first place (for Mother Nature keeps 30 trillion cancer doses of radioactivity in random places under the US). On the other hand, the amount of wastes generated per person per year by coal-fired plants amounts to 320 lbs of ash and other poisons, of which as much as 10% is spewed into the atmosphere, causing thousands of death by cancer, and by heart, lung and other diseases. The poisons produced by nuclear plants will be with us for centuries; but the poisons produced by fossil-burning plants will be with us forever.

Let us look at the details.

WHEN the uranium in a nuclear fuel rod has been spent (after about one year of service), it remains radioactive due to the intense radioactivity of the fission products. The spent rods are then immersed in deep pools of cooling water at the plant site for a few months to allow the high-level, short-lived radioactivity to die down. The rods still contain some unspent uranium, and also plutonium, which is a valuable fission product capable of being used as further fuel. The spent rods are therefore shipped in lead casks to fuel reprocessing facilities, which separate out the uranium and plutonium chemically. (The process starts by dissolving the rods in a nitric acid bath — at this stage, there is no way the mythical terrorists could get at the plutonium inside even if the rods were not intensively radioactive.) A single reprocessing center can handle as much as 5 tons of fuel per day, which corresponds to the output of eighty 1,000 MW reactors. The uranium is enriched and recycled into new fuel. The plutonium (or rather plutonium oxide) will one day be used as pure fuel in reactors which are not yet used commercially; it can also be used in "mixed oxide" fuel rods, in which a mixture of uranium oxide and plutonium oxide is used as the fuel in the rod, thus preventing the plutonium from ever achieving the ratio of volume to surface necessary (but not sufficient) for a nuclear explosion.

The remainder also contains some plutonium, since it is not possible to separate it out completely. NRC regulations require that these remaining wastes be converted to solid form (eliminating the danger of liquid spills) within 5 years after arrival at the reprocessing site, and that they be shipped to a permanent repository 10 years after reprocessing.

That is the way it should be, but in fact there is a temporary flaw in the process. At the time of writing, there is an acute shortage of reprocessing capacity, due, in part, to the reluctance of private industry to take over the reprocessing plants (hitherto run by the government) in the present uncertain climate surrounding nuclear technology, but in part also due to the issue of plutonium, which has prompted the NRC to delay recycling until plutonium safeguards have been debated and worked out. The net result is that nuclear wastes are now piling up at power plants, which are running out of space for them, and instead of being reprocessed and disposed of, they are kept in places where they might indeed become dangerous. As in so many other cases, the alarmists have not only cried "Wolf!" but they have brought in a wolf of their own.

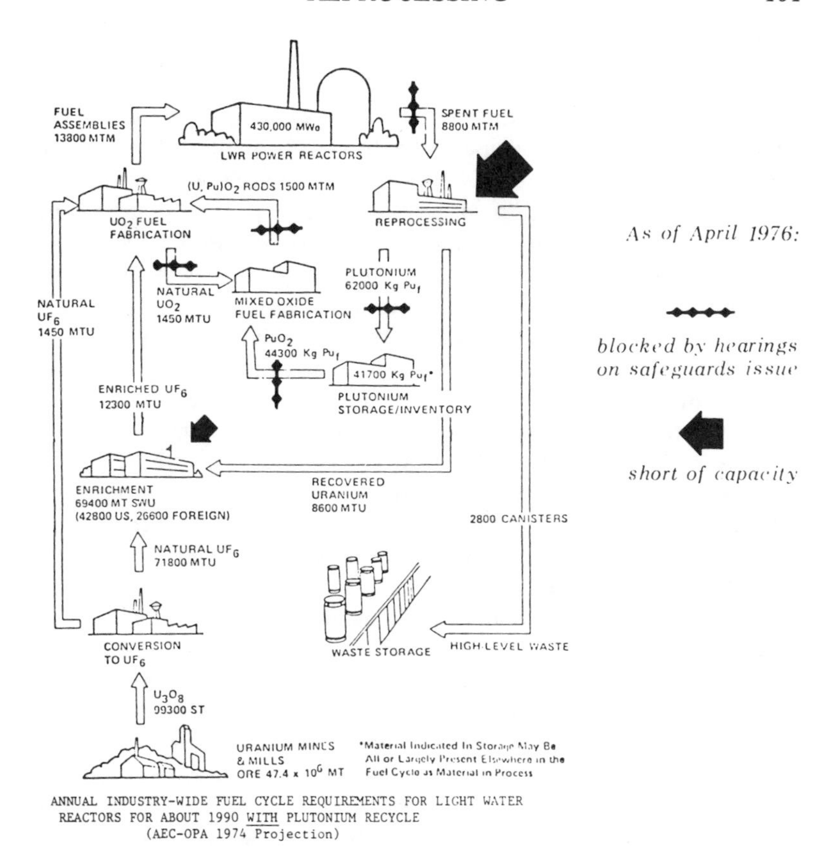

The wastes piling up at nuclear plants are a political, not a technological problem. Delaying tactics by the "environmentalists" are impeding the cycle.

However, the question of reprocessing is simply a legal one, and a temporary one at that, for reprocessing of nuclear wastes is a well proven technology that has been in use for some 30 years in the defense industry, and in the following we disregard the point, proceeding directly to the question of what to do with the remaining wastes in solid, glass-like form. These ultimate wastes are "hot" not only in the sense of radioactive, but also in the thermal sense, for their radioactivity is ultimately dissipated as heat.

These wastes present a danger only if they get into water or some other vehicle by which to enter the human body, and though nobody can give a guarantee that this will never happen, it is far easier to prevent it from happening than it is in the case of radioactive deposits that are underground naturally, not in carefully selected and monitored places, but utterly at random.

The much used rhetoric about wastes remaining "radioactive for thousands of years," while perfectly true (the halflife of plutonium 239 is 24,400 years), is quite misleading and largely meaningless. As we know from Chapter 2 (p. 54), the longer the halflife of an isotope, the less intense its radiation. Arsenic, which is not radioactive at all, has an infinite halflife, and indeed, while plutonium will be around for a long time, arsenic will be around forever.

Nor is the point about arsenic (for example) a cheap trick of demagoguery. As Prof. B. Cohen of the University of Pittsburgh has pointed out,[3] arsenic trioxide is a poison used as a pesticide. It is not a very commonly used one, but more of it (in weight) is imported every year than all the nuclear wastes would amount to if all US power were nuclear. Arsenic trioxide is about 50 times more toxic than plutonium when ingested (for plutonium being "the most toxic substance known to man" is more melodramatic piffle), but the main difference compared with the threat of wastes is this: Nuclear wastes, when there are enough of them, will be buried deep underground in carefully chosen geological formations. But the arsenic trioxide is dispersed in random places on the earth's surface, mainly where food is grown. Long after the nuclear wastes have decayed to negligible levels, it will still be around in the biosphere.

In all this scare talk about what to do with nuclear wastes without endangering future generations (we will return to that point presently), the main characteristic of nuclear wastes gets lost: Their amount is minuscule. As already mentioned, the volume produced by one person's annual share of the total US output is that of a single aspirin tablet; and this is one of the aspects that make nuclear power so attractive — not in spite of its waste disposal, but because of it. If the entire US electrical capacity were nuclear and ran at the present rate for 350 years, the volume of wastes would amount to a cube 200 feet on a side.[2] After three and a half centuries! (Actually somewhat more space would be needed in practice for cooling passages and accessibility.)

There are several satisfactory methods of disposing of the final nuclear wastes, but if a decision were made by the NRC tomorrow, there wouldn't be enough wastes to implement it. We have some two or three decades before deciding whether there are even better ways than are now being considered, whether the wastes should be disposed of permanently, or whether they should be made retrievable.

Apart from a number of highly exotic proposals (which are quite unnecessary), the simplest and evidently most satisfactory is to bury the wastes deep underground, where the chance of them ever being reached by ground water is minimal. The obvious place are salt formations, partly because the salt is evidence that no water has ever been there for at least the last two hundred million years, partly because salt formations are self-sealing in the event of an earthquake. Is it thinkable to make nuclear waste disposable safer than that?

Yes, it is. British scientists have developed a method of sealing wastes into a highly durable glass, making them fireproof, waterproof and earthquake-proof for many centuries.

A similar method of sealing nuclear wastes into glass was announced at the American Chemical Society's Centennial Conference in New York City in April 1976.

If deep burial in salt formations, after sealing in fireproof, waterproof and earthquake-proof glass makes nuclear waste disposal an unsolved problem, what, pray, is a solved problem? The disposal of fossil wastes in people's lungs?

Investigations of salt formations in Kansas proved unsuccessful, because nearby drilling holes were discovered that might have leaked water, but ERDA is now investigating other salt formations in New Mexico. There are about 50,000 square miles of salt formations in the US, so that there is no lack of sites, and burial in salt formations is only one of several viable possibilities: ERDA is planning site surveys for three facilities in each of four geologic formations — thick-bedded salt, salt domes, shale, and hard-rock (granite). It is, in any case, reassuring that the problem is being investigated decades ahead of its time.

Of course, as in all other aspects of nuclear power, nobody can give a guarantee that the wastes will be disposed of with *perfect* safety so as never to endanger anybody at all; one can only say that the danger is incomparably smaller than the presently used methods of waste disposal in coal-fired plants, and that the probability of a casualty from nuclear waste disposal is extremely small.

Moreover, as in the case of nuclear accidents, if the improbable happened and the wastes did somehow get into the ground water, the resulting casualties would not constitute the disaster depicted by emotional critics (such as Hannes Alfven who has talked about "poisoning the entire globe"). The journey through aquifers would take at least decades (allowing monitoring and countermeasures) and during that time the poisons would not only lose some of their toxicity (radioactivity) by decay, but they would also be strongly diluted. To drive this point home, Prof. Cohen made a calculation for a case which is vastly exaggerated, but allows close comparison with situations where numbers are available.[3]

Cohen assumed that the wastes from a fully nuclear US electrical capacity were to be buried at a depth of 2,000 feet, *utterly at random* — perhaps under children's schools, water supplies or any other place where blind chance happened to put them. The result of Cohen's calculation, which is based on what natural radioactive deposits are known to do, gives the expected (average) number of eventual deaths per year: 1.1 deaths for the first 200 years, declining to 0.4 deaths thereafter. And this, I repeat, under the intentionally absurd assumption that the wastes will not be buried at carefully selected sites and monitored, but that they will be buried in utterly random locations all over the US. It also assumes that no cure for cancer (the only possible hazard of wastes) will be found in the future.

AND yet, and yet, and yet... People just don't like the idea of radioactive wastes being put out of the way for thousands of years, and the reason is one of the many quirks of human psychology: They fear this danger not because it is great, but because it is *new*. They are used to millions losing their lives in wars, to thousands losing them in famines, and to hundreds losing them in railroad and airplane disasters, mine explosions, floods or hurricanes. But radioactive poisons underground, threatening somehow to get into your food — no matter how absurdly small the probability, it's new, it's a danger that wasn't there before!

The hell it wasn't. There are some 30 *trillion* cancer doses under the surface of the United States — the deposits of uranium and its daughters. They are not sealed into glass, they are not in salt formations, they are not deliberately put where it is safest; they occur in random places where Mother Nature decided to put them. And they do occasionally get into water and food, and they do occasionally kill people.

From the large amount of information on the presence of various radioactive isotopes in different parts of the human body (available from autopsies), plus some other well known information, Dr. Cohen found the number of people who die from these natural deposits:

The mean number of Americans killed by ingesting uranium or its daughters from natural sources is 12 per year.[3]

And the halflife of uranium 238 is 4.51 billion years; U 235 has a halflife of 710 million years. There is, then, nothing new about the problem; man-made deposits need not be as sloppy or as dangerous as natural ones.

"There is nothing we can do about those 30 trillion cancer doses," some people say when they first learn about them, "but at least we need not add any more to them."

But we add nothing. We take uranium ore out of the unsafe places where Nature put them, and after we extract some of its energy, we put the wastes back in a safer place than before, though we do put them back in fewer places in more concentrated form.

How concentrated? Within 10 years, more than 99.9% of the original radioactivity of the wastes disappears by decay, and the majority of the waste products then has a halflife of 30 years. In 1,000 years, the wastes are less radioactive than pitchblende (which contains 60% uranium, but also some shorter-lived and hence more intensively radiating elements such as radium). Plutonium, with its halflife of almost 25,000 years, slows the decay process, but it remains there only as an impurity that failed to be recovered for further use as a valuable fuel. And what if the Luddites have their way and dispose of the plutonium unused? Like the proverbial man who killed his parents and then demanded the Court's mercy on the grounds that he was an orphan, they want to waste plutonium and then scare people with the long halflife of nuclear wastes.

"But how can you go into something dangerous without any practical experience?" Yes, that is a problem that faced Christopher Columbus and the Wright brothers; but it so happens that it doesn't apply to nuclear wastes.

1.8 billion years before Alfven fantasized about "poisoning the entire globe" and Barry Commoner invoked images of a "nuclear priesthood watching over wastes for thousands of years," there was a natural reactor in operation in what is now the Republic of Gabon in Africa. Water pockets in a uranium deposit acted as neutron traps, and at least four, perhaps as many as six, "reactor zones" (30ft by 30 ft by 10 ft thick) went critical 1.8 billion years ago, producing an average of 20 kW thermal power for about half a million years.

The "Oklo Phenomenon" (named after the Oklo uranium mine in Gabon) was discovered when French engineers noticed a slight deficiency of the U 235 content in the ore and scientific investigations found that it had been burned up by a natural reactor. An international scientific conference, organized by the International Atomic Energy Commission, was held in Gabon in June 1975, and among the facts it firmly established were these:

There had been 12,000 lbs of fission products, and 4,000 lbs of plutonium (virtually all decayed now). All of these have remained completely in place.

In 1,800,000,000 years, 12,000 lbs of waste fission products and 4,000 lbs of plutonium have not budged an inch out of the reactor zones, though the whole phenomenon was produced by blind chance, and there were no particularly favorable chemical or other immobilization mechanisms at work.

Cannot Man do at least as well?

WASTE disposal does, however, leave unsolved problems to future generations, as well as threaten the lives and health of the present generation — when it comes to coal.

The quantity produced by one person's annual share of the output of coal-fired plants in the US is not one aspirin tablet (as in the case of the nuclear cycle with reprocessing), but 320 lbs of wastes, of which often only 90% is in the ash pile; the rest, which includes particulate and gaseous poisons, is spewed into the atmosphere, and it doesn't *threaten* to kill people; it kills them very definitely. It is these discharges into the atmosphere which are the most dangerous waste products of a coal-fired plant, but we will leave them for the chapter on routine emissions, for they are not deliberately disposed of, and hardly qualify for the name of waste "disposal."

The ash itself is not particularly dangerous, though it is not without its dangers. First of all (though this is not the biggest danger), the ash — and this comes as a surprise to many — is radioactive, too. Both "eastern" and "western" coal contains traces of radium and thorium (as well as smaller quantities of polonium and other radioactive isotopes). Nor is that radioactivity negligible — at least not compared to a nuclear plant. Its total level is higher, and it is more persistent, not only because the radium 226 in coal ash is long-lived (halflife 1,620 years), but because all radium and thorium isotopes are water soluble and chemically very active; some water-soluble radionuclides in coal ash are considered a threat to bone structure.[4] Unlike nuclear wastes, ash with its radionuclides is dispersed or buried without monitoring or control.

However, this aspect is mentioned here only under the "same yardstick" policy, for neither nuclear nor coal-fired plants add significantly to the background radiation present in the environment already, and in the case of coal-fired plants, the radioactivity (though larger than that of nuclear plants) presents a negligible risk compared to the risks associated with air pollution.

As for the coal ash not dispersed into the atmosphere, but disposed of in landfills, the real problem lies in its huge quantity.

For a nuclear plant of 1,000 MW capacity, the annual amount of solid discharges can be taken away in 60 truckloads, and even that small figure is misleading, for the heavy and bulky items of the load are the (reusable) leaden casks in which the spent fuel rods are taken away; if it were possible to load the spent fuel only, a single truckload per year would do it. On the other hand, if the 1,000 MW plant is coal-fired, the annual amount of ash taken from the plant to the dump amounts to no less than 36,500 truckloads.

The tens of millions of tons of ash generated by US coal-fired plants every year are dumped in landfills. There is enough coal in the US to last for at least two more centuries at the present rate of usage. But for how long is there enough space where to dump the wastes?

Let future generations worry.

There are no provisions to prevent the poisons in coal ash being leached out by rainwater (they are dumped close to the surface) and creeping into aquifers. The metals in it (selenium, mercury, vanadium and others) do not, like plutonium, have a halflife of 24,360 years; their halflife is infinite. There are carcinogenic (cancer-producing) hydrocarbons, such as benzopyrene, among the poisons. How many other carcinogens does the ash contain? How many mutagens (substances causing mutations) are among them?

Let future generations worry.

The radioactivity of the radium and thorium isotopes in coal ash exposes the public to at least 180 times the dose received from nuclear plants of equal capacity[5] and would violate NRC standards if the NRC were responsible for coal-fired plants, but it isn't. The radionuclides contained in coal ash are chemically active and soluble in water; yet the stuff is dumped close to the surface without strict control and without even any monitoring. Will that be dangerous in coming decades or centuries?

Let future generations worry.

volume 31 no. 1 january 1976

There is no known way of protecting people from the vast amount of wastes generated by fossil-fired plants, but there are several ways of removing the minuscule amounts of nuclear wastes from the biosphere altogether. "And getting rid of the wastes is something else again," says the little figure in the Sierra Club cartoon above. The Sierra Club endorsed a nuclear moratorium in 1974 and its executive director Michael McCloskey declared coal environmentally preferable to nuclear power.[7]

The responsibility of the present generation, it would seem from the publications of Public Citizen, Inc., is to grow hysterical about the one method of waste disposal that is capable of preventing virtually all dangers to future generations.

I have no reason to believe that the present method of coal ash disposal into landfills presents a particularly acute threat to public health, and I am not very worried; but neither are those who get hysterical about the aspirin-tablet equivalents of nuclear wastes to be sealed in glass, deposited in metal cans, thousands of feet below the surface, in salt formations, and continuously monitored — in 10 or 20 years from now when the problem will first arise.

The type of landfill depicted in this cartoon (*Sacramento Bee*, reprinted by *Critical Mass*) is, in fact, used for coal ash disposal, which is neither controlled nor monitored and leaves unsolved problems for future generations. The volume of the high-level nuclear wastes generated in the US up to the year 2,000 will amount to a cube less than 70 ft on a side. (This assumes reprocessing of spent fuel, which reduces the volume by a factor of 4.)

But once again, I neither advocate nor oppose increased concern about coal ash disposal, for my plea is more modest: Use the same yardstick.

THE logical counterpart to nuclear waste disposal is waste disposal into the atmosphere (i.e., air pollution) by fossil-burning plants. Ineptly or not, I have left this to the next chapter on routine emissions; but there is one more point about solid waste disposal, and that is that the entire problem of waste disposal from coal-fired plants is about to be significantly complicated by the use of scrubbers. In the emotional climate of the early seventies, the National Environmental Protection Act was passed in 1970, and its air quality standards are not only unrealistic (necessitating repeated postponement of their implementation), but often based on insufficient, and in some cases incorrect, data. The Environmental Protection Agency, under strong political pressures, was required to introduce quick and half-baked measures. In the case of automobile emissions, the EPA bludgeoned the auto industry into the catalytic converter, which produces sulfuric acid fumes, a health hazard that was not present in automobile emissions before, and there is also some evidence that its high operating temperatures present a fire hazard when a car is idling over inflammable material such as dry grass.

In the case of coal-fired plants, the EPA has similarly tried to bludgeon the utilities into limestone scrubbers, which are not only costly, but also partly ineffective. Some utilities, particularly the American Electric Power System ("We have more coal than they have oil"), have resisted, but most others have given in to achieve peace, if not clean air.

The logical way to eliminate the poisons put into the air by burning coal is to get rid of them before it is burned, particularly by desulfurizing the coal, or using (Western) coal with very small sulfur content. Desulfurization of coal is possible by several laboratory methods, but an economic method that works on a large scale has yet to be found. Gasification and liquefication of coal, quite apart from the economic obstacles of enormous investments, may not necessarily guarantee clean air, for it may introduce a number of other substances with which there is little experience.[6]

As for western coal low in sulfur content, the environmentalists (if they deserve that name) have been doing their utmost to prevent its use by doggedly opposing its mining.

The scrubber, therefore, makes use of the dubious and only partially effective method of polluting first and attempting to get rid of the pollution afterwards. It attempts to eliminate sulfurous pollutants (the most easily observable, but not necessarily the most dangerous) by passing the flue gases through a water spray and reacting the sulfurous compounds with limestone. A scrubber of this type is, of course, much better than no pollution control at all, but while it does not eliminate the poisonous fumes completely, it does give rise to a huge quantity of sludge, which is itself a pollutant and must somehow be disposed of.

How? That is another problem for future generations as well as the present one, for no one has yet any idea what to do with all that sludge as it accumulates.

Dr. P. Abelson, editor of *Science*, considered the point in an editorial in September 1975. He pointed out that the desulfurization systems favored by EPA are costly and unreliable, and that they give rise to a soupy sludge in vast quantities.

"If EPA standards were to be met for all new stationary sources," he wrote, "the production of sludge would rise to 300 million tons a year." In 20 years, the sludge would form a body 10 feet deep covering an area of 240,000 acres.

The editorial provoked the wrath of EPA head Russell E. Train, who fired back an angry letter twice as long as the original editorial, charging that Abelson had grossly overestimated the amount of sludge and that only 120 million tons of it would be produced per year.

That means, as elementary arithmetic will show, the sludge on those 240,000 acres will in 20 years' time be only 4 feet deep, and I hope the thought comforts you; but it leaves me very cold.

I vote for the aspirin tablets.

L-410 — "NUKES" flyer, by Environmental Action of Colorado.

5

Routine Emissions

There is more radioactive material in a reactor than 2,000 Hiroshima-sized bombs.

Ralph Nader[1]

Stating that these materials are present in a reactor, if there is no bomb to spread them over an area, is scaremongering. It is equivalent to saying that the chlorine gas stored at the city waterworks and swimming pools is sufficient to poison everyone in the city 8,726 times.

Dr. R. Philip Hammond[2]

A nuclear power plant causes no air pollution, and its only emission is radioactivity, which is quite negligible compared with the radioactive background due to either natural or (non-nuclear) man-made sources; it is also smaller than the radioactive emissions by a coal fired plant.

To repeat some of the figures from Chapter 2, the specific data are as follows: According to the Environmental Protection Agency, the average US resident gets a dose of 0.01 millirems/year from all the nuclear power plants in the country; the NRC allows 10 mrems per year to persons living next to the property line of a nuclear plant, but its guidelines recommend a maximum of 5 mrems/year, and in point of fact, it starts investigating when this guideline limit is even approached.

This question was posed by the *Mother Earth News*. If your sister feels endangered, she can always move away; but she should be careful that her new drive to work is not 0.1 (one tenth!) of a mile longer. If it is, the statistics are against her.

In comparison, a person receives an internal dose of about 20 mrems/year from his own blood (mainly due to potassium 40, contained in many protein foods), 35 mrems/year from building materials, 35 from cosmic rays, 25 from food, 11 from the ground, 5 from the air, 103 from X-ray diagnostics; the total average background dose in the US is 248 mrems/year.[3]

The unfounded fear of low-level radioactivity is often amusing. For example, Coloradans for Safe Power, an organization working in effect *against* safe power, spends much of its efforts on scare tactics based on the dangers of radioactivity. "No matter what the natural background is," wrote one of their members in a letter to a local daily,[4] "it is no justification for the additional radioactivity introduced by a nuclear plant." That, of course, is a value judgement with which one cannot quarrel; but it is puzzling why the writer, who is worried about the additional 0.01 mrems/year that the nearest new nuclear plant would give her, does not avail herself of the opportunity to reduce her annual dose by many thousands of times that amount — by moving out of Colorado to some place at lower elevation. (Cosmic rays give about 35 mrems/year and the amount roughly doubles with every mile of altitude.)

COAL-FIRED plants give rise to radioactive enissions, too, due to the radium and thorium (often also polonium and other radioactive isotopes) in the coal. The radionuclides are dispersed into the atmosphere via the stack, and, as noted in the preceding chapter, taken to landfills in the coal ash. It is not easy to compare the radioactive emissions of a coal-fired plant to those of a nuclear plant, because of the large variance in fuel composition, efficiency of ash collection and similar factors, but above all because coal-fired plants are not regularly monitored for radioactive emissions, nor are they subject to NRC regulations (they would violate them if they were). Estimates therefore vary widely, e.g., Lave and Freeburg[5] quote work by other researchers, according to whom the radioactive releases through the stack of a coal-burning plant pose 410 times the threat of radioactive emissions of a pressurized water reactor. Compared to other investigators, this estimate seems on the high side; however, most investigators agree that the radioactive emissions of a coal-fired plant are generally more significant than those of PWR's. We shall not pursue the point any further, since even 410 times the radioactivity of a nuclear reactor is very little to worry about, and indeed, nobody seems to be worried about radioactivity when it is generated by coal rather than nuclear fuel.

There is, in a way, good reason for such carefreeness, for whatever the hazards of the radioactivity generated by a coal-fired plant may be, they are utterly negligible compared with those of the air pollution they generate.

One of the marked differences betweem nuclear and fossil-fired power is that more than $1 billion has been spent on researching the safety of the former, so that today we know the probability of contracting cancer due to a given exposure to radiation reliably to at least 3 decimal places; but when it comes to the health effects of air pollution, the estimates of even the early deaths vary by factors of 2 to 5. Most often, the effects have to be inferred indirectly; in other cases they are simply unknown. We know, for example, the sulfur dioxide level and the meteorological conditions that prevailed in Greater London in December 1952, when 3,900 excess deaths were recorded in a single week, and such figures are used indirectly for estimates under other conditions. But we have no idea how many

Londoners have been, are, and will be dying of cancer and other delayed diseases contracted in the air pollution of December 1952.

I do not want to use Naderite scare tactics, and therefore I hasten to add that the London 1952 example is meant to indicate where our knowledge comes from; it is not meant to imply that the conditions of the 50's persist in London or the rest of the industrialized world. They don't; air pollution has been cut very significantly throughout the industrialized world in the last two decades, though it has only been curbed, not eliminated. Indeed, there is little hope of eliminating it by combatting it after it has been created; the only hope is to eliminate its source, and nuclear power is one of the few viable options of preventing rather than curing air pollution.

But there is another significant aspect to the point, while we are digressing from the simple figures, and that is that the abatement in air pollution has been achieved by introducing more and better technology, not by curbing it. If the cure prescribed by the contemporary alleged environmentalists were followed, and technology were curbed rather than augmented and improved, then the catastrophes of the 50's (which were not limited to London) would now be far more frequent. Today's air is cleaner in spite of, not thanks to, the technophobic efforts of the Friends of the Earth or the Sierra Club. Even now they are opposing nuclear power, though by air pollution or any other measure it is the safest form of power available.

But let us return to the facts and figures of air pollution. A coal-fired plant spews all kinds of poison into the air, even when it has been fitted with scrubbers and other pollution control equipment, for none of these gadgets can catch all the pollutants; they curb pollution, but they do not eliminate it. A coal-fired plant produces particulates, sulfur dioxides, nitrogen oxides, trace metals, and other pollutants.

Particulates are (or should be) caught by electrostatic precipitators just before the flue gases go into the stack of a power plant. The gases are passed through a strong electric field in which the particles acquire a charge and are attracted to the high potential of the precipitator, usually in the form of a metal rod. When they have accumulated into a thick cluster, they are shaken off the precipitator by mechanical impact and fall into the waste collection space below.

In technical terms, precipitators can be pretty good, for they can catch up to 99.8% of the particulates *by weight*. But that does not at all mean that they prevent 99.8% of the health effects caused by particulates. Far from it: The big chunks of fly ash and soot may be a nuisance, but their health hazards are probably minor. The really dangerous matter are the tiny particulates that penetrate deep into the lungs; they get past the natural filtering mechanisms in the body, and they get past the precipitators — the ones so good at filtering out the big and harmless particles that they earn the label "99.8% efficient."

It is known that these small particulates are harmful, that they cause chronic bronchial and lung diseases, including, very probably, lung cancer. But exact figures are not known; there are not even any highly reliable methods of measuring the amount of particulates at the top of the stack. The national air quality standards (primary — human health) prescribe an annual average less than 75 micrograms per cubic meter, and a maximum 24-hour concentration of less than 260 micrograms per cubic meter, but few of the large US cities are able to comply with these standards. One method widely used in other countries, including Britain and Japan, is to require power plants to have very tall stacks. This does not eliminate particulates (or other pollutants), but at least it disperses them and prevents high concentrations, which is particularly useful when a temperature inversion hovers over a city, trapping pollutants in the city air. If the stacks are high enough to penetrate the inversion, as they often are, the danger of this trapping mechanism is eliminated.

But tall stacks have been ruled unacceptable in the US. The 1970 Clean Air Act was passed in an emotional atmosphere without sufficient technical data and even less regard for realistic limitations. It was so written that courts later refused to accept tall stacks as a genuine method of pollution control (the suits were filed by "environmentalists" who probably thought they were on the side of clean air), and the result is that particulate pollution remains an ugly reminder of what happens when legislation is passed by vote-pleasing politicians who are more interested in posing as corporation-baiters than in clean air.

Sulfur dioxide pollution is the one that has been most intensively studied, and there is a clear correlation between sulfur dioxide concentration and excess deaths by lung, bronchial, heart and other diseases. This is shown, for example, by the figures in Table I on the opposite page.

We note in passing that a disaster like the 3,900 early deaths (the delayed deaths are unknown) in London is virtually impossible for a nuclear accident: The largest consequence considered in the Ras-

Table I. Sulfur dioxide concentration versus death rate[11]

Time	*Place*	*SO2 level [ppm]*	*Excess deaths*
Dec. 1952	London	1.5	3,900
Nov. 1952	New York	0.2	360
Jan. 1956	London	0.51	1,000
Jan. 1959	London	0.2	200
Dec. 1962	London	1.0	850
Dec. 1962	Osaka	0.1	60
Nov. 1966	New York	0.51	168

mussen Report was 3,000 deaths, with the philosophical probability of one per billion years per plant.

The table above gives a few data from a long list available from many cities in the US, Britain, France, Japan, Norway, and other countries. But all that follows from tables like these is that sulfur dioxide concentration is a good *indicator* of the health effects of air pollution; it does not in itself follow that the sulfur oxides are their main cause. Correlation is not necessarily a cause-effect relationship. It is, for example, perfectly true that "The more churches in a city, the more crimes are committed there," for the simple reason that bigger cities tend to have more churches and they also tend to have more crimes than smaller communities, because both are (roughly) proportional to the number of inhabitants. There is a definite correlation, but obviously one is not the cause of the other; one cannot reduce the crime rate by tearing down the churches.

Similarly, the concentration of sulfur dioxide is a good *indicator* of air pollution, and if the number of excess deaths is compared with it (as in the table above), there appears a definite tendency for the two variables to go up and down together. Moreover, we know that sulfur dioxide is indeed the cause of *some* adverse health effects — it is, for example, responsible for forming sulfuric acid vapor in the atmosphere, which in turn causes bronchial diseases and aggravates others. However, we do not know for certain whether, besides being a good indicator, it is also the main cause of deaths and diseases caused by air pollution.

Much less than even that is known about nitrogen oxides, which arise in combustion at high temperatures from the nitrogen in the air. The main culprit is the automobile engine, but fossil-fired power plants produce them, too, and health authorities are growing more and more disturbed about them.

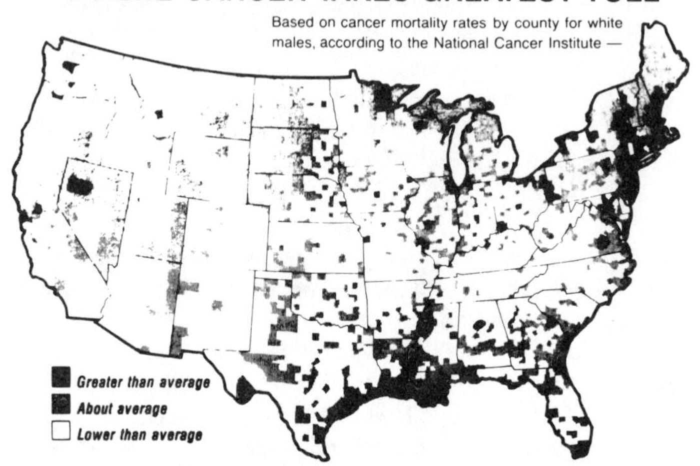

Evidence that cancer may largely be a man-made disease: Incidence correlates well with industrial concentrations. Together with automobile emissions and chemical industries, fossil-fired plants are now among the prime suspects (see text). Among the few industries clear of suspicion: nuclear.

Like other air pollutants, nitrous oxides are highly correlated with the incidence of heart, lung and bronchial diseases, but what has scientists particularly worried at the moment is that they might well be the link in a chain that could explain the high rate of cancer in cities. Cancer remains a mysterious disease, but on the basis of accumulated evidence it is now widely believed that it is largely a man-made disease whose causes are linked to the environment. Even a quick glance at a map of cancer incidence rates in the US will show the high concentrations occurring in the industrial areas.

It has been known for several years that there is a high correlation betwen the levels of nitrogen dioxide in the atmosphere and the incidence of cancer. The puzzle was that neither nitrogen dioxide nor nitric oxides (for which the former is an indicator) have been demonstrated to cause cancer themselves. But in the summer of 1975 a class of substances called nitrosamines were discovered in urban air, soil, water and sewage, and nitrosamines are a well established class of carcinogens; they are present in cigarette smoke and in some foods (such as cooked bacon, where their level is strictly limited by USDA and FDA standards). Their presence in the urban environment was a new discovery, made possible by faster and more sensitive measuring devices. The amounts measured showed that in several urban areas, breathing the air for 24 hours will deliver more nitrosamines (in the case of New York, 10 times more) than smoking an entire pack of cigarettes.[6]

The prime suspect are nitrous oxides, which under certain conditions can combine with water to form nitrous acid; this then combines with amines (organic compounds) to form the carcinogenic nitrosamines.

And where do the nitrous oxides come from? In the first place, from automobiles; but in the second place, from fossil-fired plants.

Nor are nitrosamines the only suspects; there are, in fact, not just suspects, but "convicted" culprits, such as benzopyrene, which are definitely known to be carcinogens, and definitely known to come out of stacks of fossil-fired power plants[7] (though this is not their only source).

Proceeding to other pollutants, we find even larger gaps of knowledge. For example, little is known about the actual health effects caused by metal vapors discharged into the atmosphere by burning

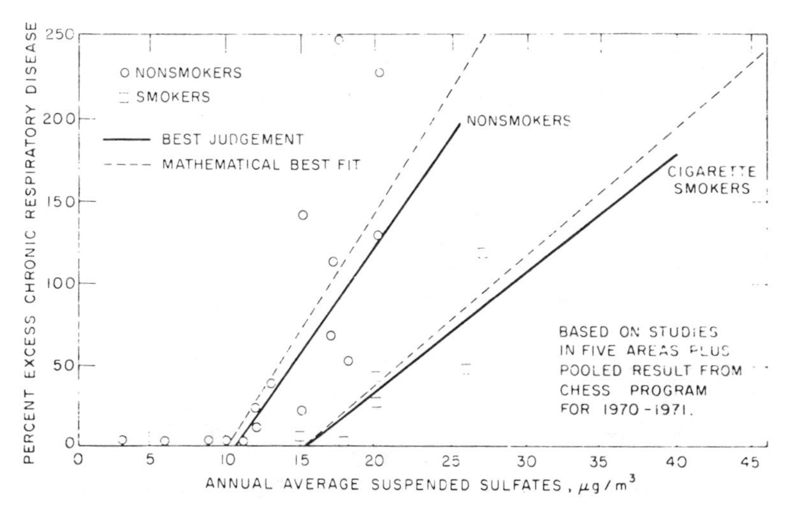

Respiratory diseases due to air pollution. (Cigarette smoking is not, of course, a protection against disease; the line for smokers is lower because lungs polluted by cigarette smoke are less susceptible to *additional* air pollution.)

CHESS stands for EPA's Community Health and Environmental Surveillance System, whose report on sulfur dioxide emissions has been sharply criticized for systematic distortion, doctored data and willful ommissions.[10] However, even large deviations from the curves above will not alter the fact that nuclear power causes no air pollution at all.

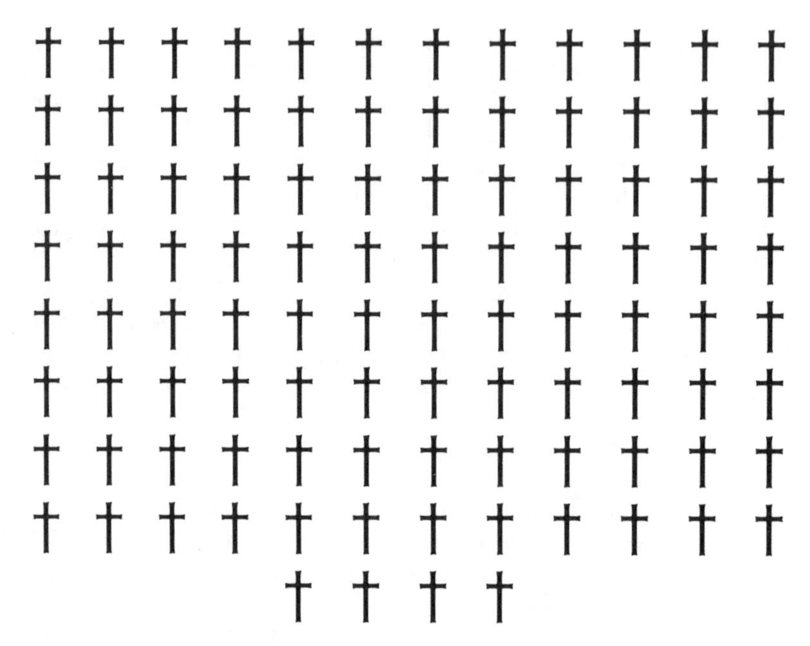

Excess deaths per year by respiratory diseases due to air pollution by US power plants: most optimistic estimate[9] and exclusive of delayed diseases such as lung cancer. Each cross represents 100 deaths.

coal, for coal contains traces of several metals, including highly poisonous mercury.

All of these emissions cause a number of lung diseases, bronchial diseases and heart diseases, often resulting in *early* death (i.e., death occurring soon after a high-pollution "episode"); how many *delayed* deaths there are, e.g., by cancer, with a latency period of up to 40 years, nobody knows.

Now all this is sharply different from the case of nuclear power, where the health effects and risks are known exactly. In part, of course, this is due to nuclear power being a much simpler danger — the only danger is radioactivity, and the only significant health hazard is cancer. But there is also another factor which must have made its contribution: More than $1 billion has been spent on studying nuclear safety. No comparable effort has been made to understand the deadly effects of burning coal and other fossil fuels.

It is, however, possible to measure the effect of burning coal (or other fossil fuels) by plotting the value of an indicator of air pollution, usually the concentration of sulfur dioxide or suspended sul-

Nuclear power saves lives: between 800 and 4,000 lives a year at present, and between 20 and 100 additional lives for every 1,000 MW nuclear plant that will replace a coal-fired plant in the future. That is also the amount of lives sacrificed by each year of delay in constructing a 1,000 MW nuclear plant. For that case, each cross above represents between 1 and 5 such sacrificed lives.

fates, against the number of excess deaths or chronic diseases, and then to fit a straight line through the cloud of points thus obtained. The straight line will then give an approximate dependence, even though the actual detailed mechanism (what pollution component causes what type of disease) remains unknown. An example of this "epidemiologic" technique is shown on p. 119.[8]

It will be seen from these data that air pollution has a "threshold," i.e., a value below which it does not cause any excess deaths or diseases. The underlying reason for this is not altogether clear. It could be that below this value pollution is harmless because, for example, the body has a chance to repair the damage caused to it by that small amount of pollution. It could also be that below the threshold the deaths and diseases are so few that they no longer appear as "excess" deaths, but disappear in the general background. But what is almost certain is that such low pollution levels are not realistically attainable in the near future.

Rose *et al.*[9] estimate the number of excess deaths due to respiratory diseases between 20 and 100 per 1,000 MW coal-fired plant per year. This is of the same order as the estimate by Rollins *et al.*,[10] between 40 and 100 excess deaths per 1,000 MW coal-fired plant per year. Using the latest available figures (1974), with 53.1% of all electricity supplied by coal-fired steam, and a total US capacity of almost 500,000 MW, *this translates to no less than between 10,000 and 50,000 excess deaths per year!*

Figures of the same order have also been calculated by Wilson and Jones,[11] and Lave and Siskin.[12]

These figures are admittedly shaky and gained by indirect inference, which is why an entire range rather than a single figure is given. But suppose even the entire range is off by a factor of 5; then there would be "only" between 2,000 and 10,000 excess deaths. On the other hand, suppose it is off by the same factor in the other direction — then the result lies between 50,000 and 250,000 excess deaths per year.

And once more we recall that these are only the early deaths, directly traceable to air pollution episodes. The delayed deaths, particularly cancer, are not included in these figures.

Perhaps the estimates quoted here are exaggerated and on the pessimistic side. But take the most optimistic estimate and make the rosiest assumptions: There still remains a risk to life and health compared to which routine emissions and waste disposal by nuclear plants are outright laughable.

WHAT these figures boil down to is this: Every 1,000 MW of nuclear power that replaces coal-fired power saves between 20 and 100 lives a year. The present nuclear capacity of roughly 40,000 MW is already saving between 800 and 4,000 lives every year. These are not lives in hypothetical accidents that might or might not happen, they are lives of Americans who are now among us and who would be lying dead in their graves if Ralph Nader had his way.

Conversely, every year of delay in which a nuclear power plant is *not* built to replace 1,000 MW of coal-fired power kills between 20 and 100 people. True, cancer, arterial and respiratory diseases kill a total of about 1,300,000 Americans every year,[13] and the names of a few hundred among them who were killed by lack of nuclear power are unknown. Their widows and orphans do not come to weep in the offices of Nader's Public Citizen, Inc.

But that does not make Nader's or Brower's attitude any less despicable. Having flatly declared that the nuclear power issue cannot be left to scientists but must be settled by "citizen activity," they cannot escape moral responsibility for these deaths. It would be callous enough to crusade against a technology that saves hundreds of lives every year, whatever the alleged motivation. But it is vile to crusade against it in the name of safety.

6

Environmental Impact

> *[B]y the end of the decade our rivers may have reached the boiling point; three decades more, and they may evaporate. . . One of the causes of this thermal pollution is the spread of nuclear power across the land.*
>
> Edwin Newman, "In Which We Live," NBC-TV, June 1970.

We have seen that the hazards of fossil-fired power to human health and safety are far greater than those of nuclear power; coal-fired plants kill, by air pollution alone, about 100 times as many people as all of the nuclear cycle, including its most dangerous phase, uranium mining. In accidents, minor or major, the ratio is of the order of 100 : 1, again in favor of nuclear power.

But what about the effects on the land and on nature in general? Here the comparison is again in favor of nuclear power, but by an even greater ratio.

THE difference is most striking in the volume of earth that has to be disrupted to mine uranium ore on one hand, and coal on the other. This, of course, is again a consequence of the high concentration of energy in uranium and the low concentration in coal: Vastly more coal must be mined to produce the same amount of electrical energy.

The annual US consumption of electric power is now getting close to 2 billion MW/h/year (by the latest available figures, 1974, it was 1.887 billion MWh/year). The uranium ore that must be mined to produce that amount of energy, assuming the use of breeder reactors, is only 200 feet by 200 feet by 100 feet.

But the volume of coal that must be mined to produce those 2 billion MWh/year is 200 feet by 200 feet by 100 miles![1] In other words, by mining uranium instead of coal, disruption of the earth could be reduced by a factor of *five thousand.*

Yet the Friends of the Earth have made it their official policy to oppose nuclear power, and most of their activity is now devoted to that effort. Clearly, the Friends of the Earth are no friends of the earth.

Is it unfair to invoke the breeder, which is not yet in commercial operation in the US? (It is in France and the USSR, and soon to begin in Britain.) Is it unfair to talk about volume rather than the disturbed surface area, regardless of the volume underneath?

Then take the figures used by the Council of Environmental Quality, an official government agency, and one virtually always on the side of the "environmentalists." According to the CEQ,[2] the annual environmental impact in land use for a 1,000 MW coal-fired plant (load factor 75%) is 9,120 acres if the coal is deep mined, and 14,010 acres if it is surface mined. Plus 161 acres for processing, 2,213 acres for transport, and 969 acres for conversion (includes 117 acres for ash storage, 13 acres for coal storage, and land affected by thermal discharges). About half of US coal is surface mined, so that the grand total is 14,635 acres.

For a nuclear plant with the same power and load factor, the annual acreage used in mining is 785 acres, for conversion 314 acres (2.2 times less), for processing 9.12 acres (17.6 times less), and the acreage used for transport is given as zero (evidently meaning negligible — one can load a year's nuclear fuel supply as an afterthought to a train taking in a day's supply of coal).

But even neglecting the other components, in mining alone the figures are 11,565 acres for coal vs. 785 acres for uranium ore, a ratio of 14.7 : 1 in favor of nuclear power.

And that ratio would increase to 4,420 : 1 if the uranium were used in a breeder. That must be why the Friends of the Earth are so fanatically opposed to it.

After figures like these, a discussion of oil spills, land use for oil and gas pipelines, and other such items would have the flavor of

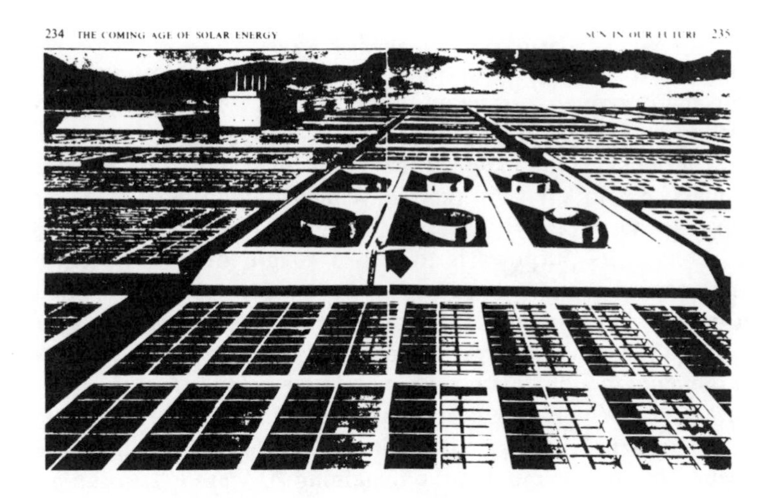

Artist's impression of a solar power plant. Arrow points to a human figure, dwarfed by its surroundings. A 1,000 MW fossil-fired or nuclear plant can be built on a few acres; a 1,000 MW solar plant would need 50 square miles of collecting area. No technological advance can ever change the dilute solar influx of 1 kW/m².

overkill, but one should mention the biggest environmental booboo of them all, solar energy. The 1,000 MW plant discussed above, whether nuclear or fossil-fired, needs about 25 acres for the plant itself plus storage facilities, rail yards, etc. A solar plant producing that amount of power (with 10% efficiency and 50% spacing between the collectors) would need 50 square miles. This has nothing to do with economics and is a simple result of the fact that solar power comes in at the dilute rate of 1 kW per square meter — at best. This would, in itself, lead to large collection areas, but since the sun is not out at night and during cloudy days, the plant would have to be designed for a much higher capacity, enabling its storage facilities to supply an average of 750 MW as above (1,000 MW times load factor 75%) when the collectors are ineffective due to the absence of strong sunlight.

Fifty square miles! The figure speaks for itself, yet I cannot resist the temptation of reminding the "environmentalists" of one of their most cherished slogans: Small is beautiful.

IT TAKES energy to produce energy: One must burn diesel oil to run the pumps that bring oil to the surface, for there are no more "gushers" in America. This has given rise to "energy accounting," a type of bookkeeping in which the debits and credits are not posted in dollars, but in kWh or other convenient units of energy. It serves various purposes, and it must be modified depending on the purpose. One of its purposes is to give a rough idea of the environmental impact of using a certain energy source; obviously, a source that needs a large amount of "debit" energy to produce only a small amount of "credit" energy is not likely to be very kind to the environment, though there is no one-to-one relationship.

For each kWh of chemical energy contained in the coal underground (1 kWh is contained in about 4 oz of coal), only 80% is recovered by surface mining, the remaining 20% are left in the mine, and 0.8% of the original energy, or 8 watts, are used in mining the coal. In processing, 7.9% of the coal is lost, and 0.1% of the incoming coal is used up in the machinery. In transporting the coal, 1% of the coal is lost, mostly by the wind blowing the coal dust off the railcars, and 0.9% is used in hauling the coal, so that more energy is lost as coal blown off the train than is used in hauling it. (Yes, that surprised me, too, but that is what the US Bureau of Mines statistics say.)

When the coal arrives at the power plant, there are only 71.3% of the original 1 kWh left, and that is now about to be really butchered: The power plant converts 38% of it to electricity, the rest is converted into waste heat; finally, 8.8% of that electricity is lost (as heat) in the transmission lines and transformers before the electricity arrives at the main switch of the consumer. The energy he starts out with (regardless of how much he, too, then turns into waste heat) is 24.9% of the energy that was originally contained in the coal lying underground.

If that sounds like very little, it is more than any other form of electric power generation (except hydropower); the corresponding system efficiencies are given in the table on the opposite page (system efficiency is the fraction of energy available at the consumer's electric meter from the total energy contained in a volume of fuel before extraction).

The fact that the net energy available to the consumer is always less than the original energy contained in the source has led many people astray and to absurd conclusions. The fallacy is that the

Fuel	*System efficiency* [%]
Coal (surface mined)	24.9
Natural gas	23.5
Coal (deep mined)	17.8
Uranium (not using breeder)	16.3
Oil (offshore)	12.9
Oil (onshore)	9.8

energy of the coal left in a mine is very different from the energy expended in hauling it to the power plant. The latter was converted and invested by man and represents a genuine loss; the energy in the coal was put there by the sun millions of years ago and is not our investment to lose. (Did I suffer a business loss when I failed to be born into the Rockefeller family?)

With the possible exception of some foods, the energy return on invested energy is always positive, or nobody in his right mind would produce it in large quantities. (Beef is eaten for its good taste, not to keep the national energy budget balanced.) The *energy gain* is the energy made available to the consumer divided by the man-converted energy invested in the production chain. In the preceding example the energy gain amounts to 16.2, i.e., the electrical energy delivered to the consumer is 16.2 times larger than the energy invested in the entire production chain from surface mine to consumer terminals.

The energy gains for other production chains (all ending with electric power to the consumer terminals) are the following:

Deep-mined coal 13.5; natural gas 4.9; nuclear 3.6; oil 2.7.[3]

Why is the energy gain so small for nuclear power, or at least much smaller than for coal? For a single reason: enrichment. More than 40% of the originally present U 235 is lost, together with its energy, in increasing the fraction of the fissile U 235 in the uranium ore, which is mostly U 238, i.e., enriching the original fraction of 0.7% to about 3.5%. The diffusion process which achieves this requires large amounts of energy to compress and pump the uranium hexafluoride gas through thousands of stages and membranes to achieve a partial separation of the two isotopes.

However, it will not always stay that way. There is not much one can do about the energy efficiency of coal mining after many centuries of its evolution. But there is real hope for making the enrichment process drastically more efficient.

Centrifugal separation is basically the same process as is used to separate cream from milk, but requiring vastly higher speeds and far greater stresses on the metal of the centrifuge. Because of the latter

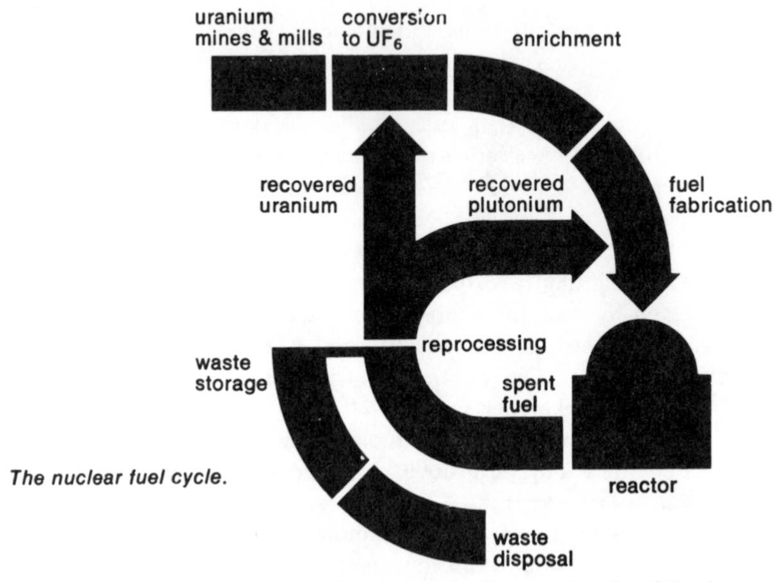

The nuclear fuel cycle.

difficulty, enrichment by diffusion was chosen in the Manhattan Project and it has remained with us ever since. However, centrifugal separation is expected to increase the efficiency of enrichment by a factor of no less than 10, and mechanical problems that were insurmountable in 1942 are now solved, though not yet tested in large-scale and prolonged practice. Centrifugal enrichment will almost certainly be used in Europe, and very probably in the US, too. When that happens, the energy gain of nuclear power will shoot past that of coal.

There is also another method under intensive investigation, though as yet only for minute quantities of uranium in the laboratory. It is based on the ionization of one of the isotopes by laser radiation, and then separating the two isotopes electromagnetically. The energy gain would then increase even more, but the method is not likely to be developed to commercial size in the next decade.

In the meantime, there is a very simple way to decrease the energy used for enrichment, and that is to avoid it altogether in using recycled plutonium oxide for mixed oxide fuel (the uranium oxide still needs enrichment as shown in the figure above). But the "environmentalists" are opposed to reprocessing because of terrorism and sabotage, which we will examine in the next chapter.

Heads, they win; tails, you lose.

By various miscalculations and misapplications of energy accounting, some opponents of nuclear power have asserted that a nuclear plant must run for half its life before it repays the energy that went into the enrichment of its fuel, and some have even charged that there is no net energy gain at all. That is patent nonsense. A 45 MW plant is enough to enrich the fuel required by a 1,000 MW plant. And about 6% of a reactor lifetime output is needed to build and operate the reactor. When thus formulated as a fraction of the lifetime output, nuclear already surpasses coal: The corresponding values are 6.7% and 7.8% for a power plant burning surface-mined and deep-mined coal, respectively.[4]

THERE remains one more ostensible type of environmental impact, so-called "thermal pollution." There are few parallel cases where the latter-day environmentalists have so exaggerated, confused and entangled an issue, and we begin by disentangling it.

An electric power plant, whether nuclear or fossil fired, converts only about one third of the fuel energy into electric energy, the remaining two thirds are converted to waste heat. This, first of all, has nothing to do with the Second Law of Thermodynamics, which says that it is impossible to perform work by cooling a body below the temperature of the coolest point of it or its surroundings. It follows from the Second Law (by a process of reasoning found in any textbook on the subject) that in every energy conversion *some* energy is irreversibly converted to heat, but it does not say that the two thirds of the energy lost in today's power plants *must* be lost as heat. In fact, the heat now lost, or at least a large part of it, could be utilized for mechanical work (in so-called "bottoming cycles"), or converted to other forms of energy, or utilized as useful heat for centrally heating homes, desalinating salt water or keeping fruit orchards warm. Alternatively, production of industrial process steam (using no less than 17% of US consumption of primary fuels) could often be modified to make steam hotter and to go first through a turbogenerator in a "topping cycle;" the electric power could be sold to utilities for distribution through their net.[5] This would greatly increase the efficiency of electric power generation (the full reasons need some familiarity with thermodynamics, but the main idea is that the heat, instead of being wasted, is exploited for the same purposes as before the topping cycle was added).[5]

As the price of fuel increases, it is very likely that some of these methods will be used; there are many reasons why they are not used now, but not one of them has anything to do with the Second Law.

Second, thermal "pollution" is most often based on colossal exaggerations, colossal even by environmentalist standards. It is true that as more and more energy is converted, more and more heat must be generated (the Second Law does come in legitimately here), so that eventually a limit might be reached for large concentrations of industry. But that day, if it ever comes, is very far away. For US consumption of energy to come even within one percent of the energy incident from the sun, every American man, woman, child, and infant on the breast would have to consume, from midnight to midnight, no less than 2 MW of power, which he could do by running 600 clothes dryers all day and all night, or cleaning his teeth twice a day with 15 million toothbrushes. And the rest of the world would still have to consume twice as much.

But no exaggeration is too great for the instant ecologists, especially when they are playing politics. Some years ago, Governor Gilligan of Ohio announced that he would "back legislation making it unlawful to increase the temperature of the water [of Lake Erie] by one degree over the natural temperature." Prof. J.J. McKetta of the University of Texas has calculated that if all the electrictity produced in the state of Ohio were used for nothing else but heating Lake Erie (whose temperature changes naturally by more than 40°F from summer to winter), the water would be heated by less than 0.3°F.[6] Please contrast this with the quotation introducing this chapter on page 123.

There may indeed be problems with excessive heat production. For example, Manhattan and the Los Angeles basin have an average temperature that is almost 4°F higher than that of their surroundings. This, of course, is not caused by power plants, but by the large concentration of human population and their activities in general. Whether this poses a health hazard, nobody knows, and the environmentalists are not losing any sleep over it, for they are interested in fish, not people.

But their concern over fish is entirely misplaced, for the fish usually love what the environmentalists have misnamed "thermal pollution." If a power plant uses the water of a nearby river or lake to cool its condensers, it will raise the temperature of the water only in the immediate neighborhood of the plant, and only by a few degrees. (The usual increase in local temperature of the water is 3°F

* * *

LOBSTER LOVERS may get a break eventually if a current study finds it feasible to start large-scale lobster farming in the warm waters discharged by New England electric power plants. Studies already indicate that lobsters grown in warm water reach marketable size in two years instead of the usual five to eight.

* * *

It has long been known that the warm water near the heat sinks of power plants is beneficial for fish; this report from the *Wall Street Journal* may confirm the trend for lobsters, too.

at a point 1000 ft from the discharge.) To speak of the "destruction of aquatic life" is another colossal exaggeration, for what sometimes happens is that one species of fish moves out, but another species, preferring the warmer water, moves in, and one might well ask these muddle-headed friends of wildlife why they would deny these fish a living.

In the late 1950's, the construction of a nuclear plant on the English river Blackwater was opposed by environmentalists on the grounds that the warm water would endanger the oyster banks lower down the estuary. But the plant was built, and nothing happened to the oysters — until the severe winter of 1962-63, when many of them froze to death and the thermal "pollution" by the plant was unable to save them.[7]

It was, in fact, nuclear plants, which often used to reject their waste heat into a nearby river in so-called "once-through cooling," that showed up the highly beneficial effects on fish: They flock to the warm water, grow approximately twice as fast, and to a bigger size than in cold water. (The generally accepted reason is that they spend more time feeding than in cold water.) So successful has thermal "pollution" been in improving the habitat of fish that several fish hatcheries in Britain and the US now use thermal "pollution" (sans nuclear plants) to grow bigger and healthier specimen faster. It is for reasons like these that Prof. McKetta has suggested replacing the term "thermal pollution" by "thermal enrichment."

The witch hunt against thermal "pollution" has mainly been directed against nuclear plants, which supposedly produce much more waste heat than fossil-fired plants. In the first place, this is untrue. The waste heat produced by a power plant can be determined from its efficiency, the ratio of electrical energy produced to the energy contained in the fuel; if a power plant has an efficiency of 40%, then

40% of the fuel energy is converted to electricity, and the remaining 60% are lost as waste heat.

The highest efficiency that has been achieved in fossil-burning plants (of very large size) is 41%; the highest efficiency for nuclear plants is achieved by the High Temperature Gas Reactor, and it equals 39%, which is very close to the fossil-plant record. But most commercial nuclear reactors in the US (at present, all but the HTGR in Fort St. Vrain, Colorado) are light water reactors, which have an efficiency of only 31%. But this is again very close to the average fossil-burning plant efficiency: The latest available data (1974) show the average fossil-to-electric conversion efficiency to be 32.53%.

There is, in fact, only one substantial difference in waste heat dissipation between fossil-burning and nuclear power plants. In a fossil-burning plant, one-third of the waste heat escapes through the stack into the atmosphere, and only the remainder offers a choice for dissipation into the atmosphere or into a body of water. The water cooling the condensers (see figure on p. 36) must itself be cooled. In "once-through" cooling, this water comes from, and is returned to, a river near the plant. It can also be pumped from, and back into, a nearby lake (or artificial cooling pond); or it can be cooled by running through a cooling tower, dissipating its heat into the atmosphere.

A nuclear plant, on the other hand, has no stack and has a choice between rejecting all of its heat into nearby water, or into the atmosphere, or into both in any desired ratio. The cheapest and most effective way, of course, is once-through cooling by a nearby river, if one is available. But EPA regulations forbid, in effect, once-through cooling for power plants built since 1970. Electric power, particularly nuclear, is cheap enough to allow environmentalists to saddle the rate payer with the expense of cooling towers, which are, in many cases, giant concrete monuments to the politics of ecological folly.

ONE final remark about the "greenhouse" theory. The carbon dioxide content of the atmosphere has been increasing over the last 100 years, and it is often assumed (without sufficient evidence) that this is due to human activity, particularly the burning of fossil fuels. This, some fear, might lead to a "greenhouse effect" by which too much solar radiation is trapped by the earth, leading to a heating of the atmosphere and a generally hotter climate. (We will not stop to discuss the details of the greenhouse effect, except to remark that it plays no significant part in heating a greenhouse.) The greenhouse effect and its dangers to the environment have been seized upon not only by panicky environmentalists, but also by some advocates of nuclear power, since only fossil-burning plants produce carbon dioxide.

Yet I will not include this possibility as one of the "health hazards of not going nuclear." The reason is that some of the premises of the theory, and all of its predicted consequences, are largely speculative and full of highly debatable points. It might very well turn out to be true, but at present there is only flimsy evidence to support it. As the preceding chapters have shown, the health hazards of non-nuclear power are so real that the vastly superior safety of nuclear power has no need of such shaky arguments.

7

Terrorism and Sabotage

These plants and associated transportation vehicles, containing deadly radioactive materials, are so vulnerable to sabotage or theft that a garrison state has to be built up to try and safeguard them... Some observers believe there will be a million people with direct and backup assignments to guard the nuclear industry by the year 2000.

Ralph Nader, countless times, e.g., speech at Syracuse University, 6 April 1975.

Among the by-products generated by fission in a nuclear reactor is plutonium 239, which is itself fissile. It is toxic when eaten, and particularly when fine particles of it are inhaled — though nowhere near as toxic as some other substances. It can also, in sufficiently large and pure quantities, provide the raw material for a nuclear bomb.

The toxicity of plutonium and the feasibility of blackmailers dispersing it in a city has been exaggerated to an incredible degree, and we shall return to this point presently.

On the other hand, the threat of nuclear weapons in the hands of terrorists is a plausible possibility and should not be underestimated, though the threat comes from a different direction than the nuclear foes would have us believe.

AS IN the preceding chapters, we will refrain from comparing risks to benefits, and we will simply compare the risks of terrorism or sabotage by means of nuclear weapons to the risks of terrorism or sabotage by means of blowing up oil and gas storage facilities or hydro-electric dams. It may surprise many readers that the consequences of such acts are comparable, but the nuclear version is far easier to guard against; and the technical difficulties of sabotage and terrorism are incomparably greater in the nuclear version, too.

The reasons why this is so will be discussed shortly, but they will be discussed only for the sake of completeness and maintaining the same approach to the problem in all its aspects, including the present one. However, it should first be stressed that the comparison in the present case, i.e., terrorism, sabotage and blackmail, is largely *irrelevant.*

In all of the previous points under discussion, we were concerned with genuine alternatives. We either go nuclear or we don't, or we go partly nuclear. For every 1,000 MW of nuclear power that replaces coal or other fuels, we save some coal miners' lives, but we endanger some uranium miners; we decrease the risk of major and minor accidents associated with fossil fuels, but we increase the risk of a nuclear accident; we decrease air pollution, but we increase (however slightly) the radioactive background. Whether we consider deaths, injuries, diseases or damage to the environment, we always face a trade-off. It so happens that nuclear power comes out very superior in the trade-off; but a trade-off is what ultimately must be faced.

However, there are no genuine trade-offs involved in terrorism, sabotage or blackmail. If for some reason nuclear power were prohibited tomorrow, it would not eliminate the risk of nuclear terrorism; it would not even significantly reduce it, for it is another Naderite myth that prohibition of nuclear power in the civilian sector only, of a single country only, can in any way alleviate, let alone eliminate, the threat.

IT IS close to impossible for a single person to steal, breed, or otherwise obtain sufficient plutonium to make a bomb. It is also highly doubtful whether a loner, somehow in possession of a sufficient quantity, could manufacture a bomb and an effective triggering mechanism. It is not beyond the realm of possibility for a group of determined and technically erudite madmen (probably ideologists) to achieve this, but it is highly improbable that, apart from the enormous technical difficulties, they would choose to do so; for there are far more effective and easier ways of killing, or threatening to kill, large numbers of people. Even if such a group decided on a nuclear weapon for the purpose (which is unlikely), the easier and more effective method would be to steal, perhaps by force, a ready-made military nuclear weapon, such as a tactical nuclear bomb. With the assumed education and determination of the group, this would present fewer difficulties than the long chain of obstacles associated with manufacturing a crude (and probably ineffective) home-made weapon.

The amount of such weapons in the US (and overseas US bases) is secret, but it is well known to be quite large; the amount of weapon-grade plutonium (not plutonium oxide fuel, badly suited for weapons) is incomparably larger than the plutonium oxide fuel pellets ever likely to be shipped from reprocessing plants to power plants, which is the one and only chance to get at it. Yet for this incomparably larger amount the safeguards have worked, and without a trace of a "garrison state," a Naderite buzzword. The "plutonium economy" is another such buzzword intended to scare people. If all the US electrical capacity were nuclear, there would still be far fewer shipments of fuel assemblies than of flea collars for dogs. Would it make sense to speak of a "flea-collar economy?"

There is, however, a third possibility, which cannot be taken so lightly, and that is the terrorist, blackmailer or other agent supported, at least in part, by a foreign government. The Palestine Liberation Organization, for example, has shown an utter disregard for human life for objectives as small as getting publicity; they are not only recognized as a legitimate political organization by the USSR, but also supplied with sophisticated weaponry by the Soviets, and largely trained by Soviet military experts. At the time of writing it seems very unlikely that the Soviets would, for reasons of their own, supply the PLO with nuclear weapons (covertly, of course), but whether such a step will be taken in the future rests entirely in the humane and trustworthy hands of the Soviet politbureau.

Nor is this the only example. West Germany is about to supply an enrichment plant to Brazil, and there will evidently be little effective control over it. A secret safeguards agreement was reached among the nuclear powers (other than China) in early 1976, but ultimately it is only up to the good will of, say, Brazil whether she will enrich her uranium to 3% for nuclear fuel or to more than 90% for nuclear weapons. "We would never dream of making a nuclear bomb," said the Brazilian foreign minister in 1975. "Unless, of course, Argentina made one first."

France is offering nuclear equipment to the Arab countries in the hope of ensuring its oil supply. It is also perfectly possible to breed plutonium from unenriched uranium ore without bothering to produce electricity in the process. (It would serve no purpose here to go into the technical details of the procedure, which is well known to experts.) There is a whole string of Third World countries close to nuclear capability, and there are not many obstacles in their way if they are determined to obtain it.

The case of India teaches two lessons. First, that it is not all that easy to make a bomb. India, unlike the Arab and other backward countries, has a highly competent scientific elite. It also had full government support in breeding plutonium from unenriched uranium ore by means of the Canadian CANDU reactor. Yet it took them 10 years to manufacture a bomb, and when they had made it, it did not work; it went off only on the second try. The other lesson, of course, is that given government support, it can be done.

The danger in all this is that it will take only one precedent to open the floodgates. It matters little whether Brazil supplies a bomb to Libya to use against Israel or whether any other of hundreds of scenarios becomes the first; once the first "little" bomb goes off, none of the members of the "nuclear club," certainly not the communist ones, will hesitate to arm their clients similarly "in defense," and the floodgates will open.

Suppose, then, that the PLO or some other terrorist organization plants a nuclear bomb in Manhattan, whether covertly supplied by the Soviets or manufactured by some Third World government; what difference will it make what percentage of US electrical capacity is nuclear and what percentage is coal-fired?

That, and not some horror story about a nut building a nuclear bomb in his garage, is the real danger of nuclear terrorism, and the risks associated with plutonium shipments in the power industry are not only tiny, but above all, irrelevant. We nevertheless discuss them briefly below, after which we will come back to what can be done about the real dangers.

TURNING to technical details, there is one scenario of sabotage that can be dismissed quickly, and that is the "shooting up" of a nuclear plant. Although even Ralph Nader has abandoned the idea that a nuclear plant can undergo a nuclear explosion, he now claims that it is possible for a saboteur to "to blow up a plant with sophisticated weaponry from a hilltop" and "rupture the entire pressure vessel" so as to release its radioactivity.[1]

The containment building is made of concrete 3½ ft thick, with the steel reinforcement mesh so tightly woven (see photo on p. 47) that vibrators must be used to force the concrete through it before it hardens.[2] That makes the walls much stronger than, for example, the roofs of the German submarine bases on the French Atlantic coast, which were bombed round the clock by the allied air forces with "blockbuster" bombs, but withstood even direct hits. But suppose this imaginary supersaboteur did have some mysterious missile that managed to blow a hole into the containment building. What next? Would he have a second missile to make the hole larger, and a third to penetrate the remaining concrete structures inside the building, and a fourth to begin working on the steel pressure vessel? Would he wait until the weather is just right so that the fruits of his labors are not dispersed harmlessly in the atmosphere? This does not yet ask all the questions, but the whole idea is too absurd to waste more space on.

PLUTONIUM is often called "the most toxic substance known to man," "toxic beyond human experience," the "fearsome fuel," and other such melodramatic nonsense.

Of course plutonium is toxic. Of course it must be handled with care. But the rest is just horror propaganda. Plutonium is primarily an alpha emitter, which means that its radiation is absorbed in the air after a few inches, and a sheet of paper is sufficient to shield oneself against its radiation at close quarters. It is far from being the most toxic substance known to man. When eaten or absorbed in the blood stream, it is ten times less toxic than lead arsenate and hundreds of thousands of times less toxic than some biological poisons such as diptheria or botulism toxin. Caffeine, some of which you probably had this morning in your coffee, is only 10 times less toxic than plutonium. (Relative toxicity is measured by comparing the weights of 50% lethal doses given to the same type of mammal. The "50% lethal dose" is the amount that will kill half of the experimental animals.)

However, though ingestion of plutonium or its absorption through the skin is dangerous, the real danger of plutonium is breathing it in the form of fine dust particles. Plutonium is insoluble in water, and fine particles may stay long in the lung, eventually causing lung cancer. Even so, this danger, which undoubtedly is a serious one, has been exaggerated beyond all reasonable bounds. There are radioactive substances produced not by the Pentagon, but by Mother Nature, which are far more dangerous than plutonium.

"Plutonium is the most toxic of all elements," goes one of the scare slogans of the anti-nuclear movement. Why of all elements? There are only about 100 of them, and they have no reason to be particularly toxic. But the devious trick doesn't even work, for the statement remains false. All the heavy metals are toxic, and several of them are more toxic than plutonium. Radium has a halflife 16 times shorter than plutonium, so that at first sight it would seem 16 times more dangerous (because for the same number of atoms, its intensity of radiation will be 16 times greater). However, plutonium has a four times longer residence time in the lung, so that it is only four times (16/4) less dangerous than radium. There are many other examples, but this one should be enough to dispel the myth.

"A few ounces of this deadly stuff, if properly distributed, could wipe out all of mankind." And so could a day's production of sewing pins, "if properly distributed" (one into every man's heart). The fact is that the amount of plutonium committed to the atmosphere in the early atmospheric tests of nuclear weapons was not a few ounces, but almost *three tons*;[3] yet somehow mankind survives.

Statements of this type are made by the Naders, Koupals and Comeys, political propagandists without any scientific training. But how about radiologists like Sternglass, Geesaman, Gofman, Tamplin and Cochran?

The only thing notable about these ex-scientists is that they get a lot of publicity every time they make one of their wild charges. They have been refuted, time and again, by scientific committees and professional organizations investigating these charges. But these rejections by large bodies of scientists and professional organizations do not make the news.

But hasn't it happened before that a scientist challenged the conventional wisdom of his time and earned the hostility of his colleagues only to be vindicated in time?

Yes, it happened to many great scientists, Galileo, Darwin, and Einstein, to name a few. But there are a number of important differences, all of them open to inspection by laymen.

First, these great scientists earned the long-lasting hostility of politicians, ideologists or religious fanatics; the scientific community was fairly quickly persuaded that they were right. For example, the

Nazis branded Einstein's theory a "Jewish hoax;" but their scientists were using it nevertheless.

Second, the discrepancies between old and new theories in cases where they could be put to experimental tests, were very small. The time difference in Galileo's experiment with the two balls was a fraction of a second, and Einstein, in 1905, had only two experimental results to go on: One was the mere absence of a small effect in a highly sensitive interferometer, the other a difference in an electronic effect so small as to be (at the time) barely detectable. That is hardly the case with the Don Quixotes of radiology. For example, while US safety standards require that a "hot" particle of plutonium be considered to deliver a dose of 0.3 mrems/year, Drs Tamplin and Cochran calculate the pertinent dose to be 4,000,000 mrems/year,[2] and while the minute discrepancies predicted by Einstein's theory provoked scientists into heated debates, a discrepancy by a factor of more than 10,000,000 will usually provoke them only to snicker and tap their foreheads.

Third, the genuine scientist who challenges conventional wisdom faces the hostility of a world that does not like to have its sacred cows slaughtered. Galileo had to revoke his statements under threat (perhaps even application) of torture. Giordano Bruno was burned at the stake. Darwin earned the life-long hostility of the Church. Einstein was driven into exile together with many non-Jewish scientists who supported his theory.

But Gofman, Tamplin & Co are in the very opposite position. Today it is the genuine and hardworking scientist whom much of the world regards as some kind of Dr Frankenstein if he supports nuclear power, and it has taken the Sternglasses and Geesamans to its bosom. They have access to the lavish funds of the various environmental foundations (and the eqully lavish funds of the para-religious foundations such as the Creative Initiative Foundation), they travel the lucrative lecture circuit, they bask in the publicity of the mass media, they can not only ride their little pet hobby horses, but they can do so while posing as prophets, martyrs and saviors rejected by a callous, profit-greedy establishment. They have, in short, discovered a shortcut to the glory that they failed to reach by consciencious and responsible hard work.

Given these criteria, it should not be difficult for the layman to decide whether the truth lies with Tamplin & Co or whether it lies with the American Health Physics Society and the Committe for Biological Effects of Ionizing Radiation of the National Academy of

THE SUNDAY OREGONIAN, AUGUST 24, 1975

Scientist links N-plants, crime

CORVALLIS (AP) — Northwest Oregon could experience a crime wave in December when the Trojan nuclear power plant at Rainier goes into operation.

That is, if one behavioral scientist's theory is accurate.

G.D. Hanks, a faculty member at Indiana University Northwest in Gary, says his research indicates a connection between violent crime and gaseous nuclear power plant emissions.

He outlined his findings just concluded at the American Institute of Biological Sciences convention at Oregon State University.

Hanks studied Federal Bureau of Investigation statistics on murder, forcible rape and aggravated assault in cities near the sites of nuclear power plants. He said he found that increases in the crimes occurred when the nuclear plants opened.

The cities he studied included Joliet, Ill., which is about 18 miles downwind from the Dresden nuclear power plant, and New London, Norwich and Hartford, Conn., cities near the Waterford nuclear power plant.

Hanks compared crime rates before and after the plants started operation. He also compared the rates of the test cities to other more distant municipalities of their comparable sizes.

There was a substantial difference in the comparisons, Hanks said. Cities near atomic plants had higher increases in murder, rape and assaults after the plants opened than did other cities the same sizes.

"We have been brainwashed that nuclear power plants produce only small emissions, if any," he said. But Federal Environmental Protection Agency scientists measured the radiation in

---

"His allegations, made in several forms, have in each instance been analyzed by scientists, physicians and bio-statisticians in the Federal government, in individual States that have been involved in his reports, and by qualified scientists in other countries. Without exception, these agencies and scientists have concluded that Dr. Sternglass' arguments are not substantiated by the data he presents. The U.S. Public Health Service, the Environmental Protection Agency, the States of New York, Pennsylvania, Michigan and Illinois have issued formal reports in rebuttal of Dr. Sternglass' arguments. We, the President and Past Presidents of the Health Physics Society, do not agree with the claim of Dr. Sternglass that he has shown that radiation exposure from nuclear power plant operations has resulted in an increase in infant mortality."

Nuclear power increases crime and infant mortality. It is extremely unusual for a professional scientific organization to rebuke a scientist, but the allegations made by Sternglass, Tamplin, Geesaman and others in the anti-nuclear group are so wild and irresponsible that several organizations felt obliged to issue public rebuttals. Above is an excerpt from a statement signed by the President and every Past President of the Health Physics Society rebutting Sternglass' claims about nuclear plants increasing infant mortality. (The Health Physics Society is devoted to the research of damage to human health and environment by radiation; it has no stake in nuclear power).

Sciences (to name but two of many institutions). But if that is not sufficient, there is always the ultimate test: the experimental evidence.

And the experimental evidence is overwhelming. Not a single human cancer has ever been positively associated with exposure to plutonium.[5] During the national emergency conditions of the early nuclear weapons industry, the exposures to plutonium far exceeded the present maximum permissible limits. Yet of 17,000 plutonium workers, including those associated with the Manhattan Project, not one has died of plutonium-related health problems.[6]

Included in this figure are 25 plutonium workers from Los Alamos (1944-45) who had *twenty-five times* the presently permissible amount of plutonium deposited in their lungs. (Tamplin and Cochran advocated a reduction of the present maximum permissible lung burden by a factor of 117,000. Their petition via their patrons, the Natural Resources Defense Council, to the NRC to this effect was under study for two years and rejected by the NRC in April 1976 on several grounds, one of which was the NRDC's misinterpretation of scientific data.) According to Gofman's estimate of lung damage, these 25 workers should have developed 1,500 individual lung cancers.[7] In fact, all twenty-five are in good health.[8]

THE THREAT of plutonium dispersal by terrorists or blackmailers has been thoroughly analyzed by Dr Bernard Cohen, a past Chairman of the American Physical Society's Nuclear Division, and any terrorist who should read his study[9] will be sharply disappointed, for plutonium is not merely far less toxic than some other poisons, but unlike chemical or biological poisons which kill their victims within a few minutes, plutonium-caused death (cancer) is delayed by years and decades.

If, for example, someone were to take Ralph Nader's irresponsible insinuations seriously and disperse plutonium into the ventilation system of a building, the victims would be left with 15 to 45 years of good health. The type of terrorist who might commit such a senseless and inept crime could be one who has been reading Naderite or similar anti-nuclear hysteria and taking its implications seriously. While Nader or Koupal would be innocent of such an event in the eyes of the law, it is difficult to see what excuse they might hide behind to escape moral responsibility.

The method could not even be used for blackmail, since the threat could immediately be defused by shutting off the power to the building and its ventilation system. Even if the blackmailers did not reveal the specific building, power to all large buildings in a city could be turned off for a comparatively short time, for plutonium differs from biological and chemical poisons in yet another way: Exceedingly small amount of it (or any other radioactive substance) can quickly be detected.

For the rest, we refer the reader to Dr Cohen's study;[9] the threat of plutonium dispersal is highly improbable, because so many other more dangerous and more effective methods are available to terrorists and blackmailers with much less trouble and danger to themselves.

THE ISSUE of illicitly manufactured bombs is another matter. It has been thoroughly discussed in *Nuclear Theft: Risks and Safeguards* by Theodore B. Taylor, a nuclear physicist, and Mason Willrich, a professor of law. Quotations from the book have been repeated innumerable times by nuclear critics, environmentalists and political foes of nuclear power. Given the vast quantities of utter nonsense spewed out by the anti-nuclear organizations, the disinformation spread by ostensible documentaries (such as "The Plutonium Connection") on TV, and the exaggerated publicity given to the discredited theories by Gofman, Tamplin and others, and given, moreover, that the study was commissioned by the Ford Foundation's Energy Policy Project (which mainly engages in ideological economics), the first reaction of many has been distrust and suspicion that the Willrich-Taylor study is more of the same propaganda.

That is not so at all. It is a serious work by highly qualified investigators who have produced a first-rate document. The quotations taken from it and indiscriminately thrown about (until eventually they become unrecognizable parodies of the original) are most often used by political activists who have never been near the book, or they would realize that it was not written to stop nuclear power, but to make it safer. There was indeed undue laxness in the security of some of the phases of the nuclear fuel cycle when the book was published (1974), and it doubtlessly played no small part in bringing about the remarkable tightening of security at nuclear facilities that has taken place in the last two years.

Indeed, Dr Taylor has himself stated that with the recent improvements and currently proposed upgrading, he believes the safeguards program will be satisfactory before significant quantities of plutonium begin to flow through the nuclear fuel cycle; [10] and Dr Willrich has testified against the California shut-down initiative, affirming that "the existing legal and regulatory framework for nuclear safety in the US can be made to function effectively." [14]

What Taylor's and Willrich's book shows quite clearly is how enormously difficult it is to get at the plutonium and then to make a crude bomb out of it. There is no way a terrorist or anybody else can get at the fission products of a reactor while it is in operation, for example. The only time an unauthorized person could get near plutonium is when fuel is shipped from the processing plant to the reactor. It will be in the form of fuel rods inside which plutonium oxide is mixed with uranium oxide. A glance at the figures on the opposite page will show how easy it is to get at them.

The plutonium oxide used in a reactor is a very different matter from the plutonium used in weapons; it would, however, be possible to make a bomb from it, though only with enormous difficulties. To steal the several tons of lead container with the sub-fuel-assemblies in it from a train or truck would have to be a major, yet covert, military operation. Transport of plutonium must now include armed guards plus an escort vehicle with armed guards who have "shoot to kill" orders. They must be in constant radio contact with outside monitors, and if they pass through areas where this is not possible, a second escort vehicle must be added. Does that have the makings of a "garrison state?" If it does, then there are some far more conspicuous examples — the anti-hijacking measures at airports, for example, against which few reasonable people have raised objections. Besides, fuel shipments to a nuclear plant ordinarily take place only about once a year; but tons and tons of plutonium have been shipped in the weapons industry for more than three decades without incident, and without the US becoming a "garrison state."

There is a host of other weighty objections to the "domestic" terrorist scenario. It assumes that a team of scientists capable and willing to produce a bomb will remain motivated without the story leaking out for the duration of several months that such an enterprise would require. [9]

More to the point, if these "mad scientists" chose the nuclear option for their designs, wouldn't it be far easier to acquire, by force or intrigue, a ready-made nuclear weapon? Or at least highly enriched uranium 235?

These scientists would have to be sufficiently logical to think up ways of producing a home-made bomb, yet sufficiently stupid to

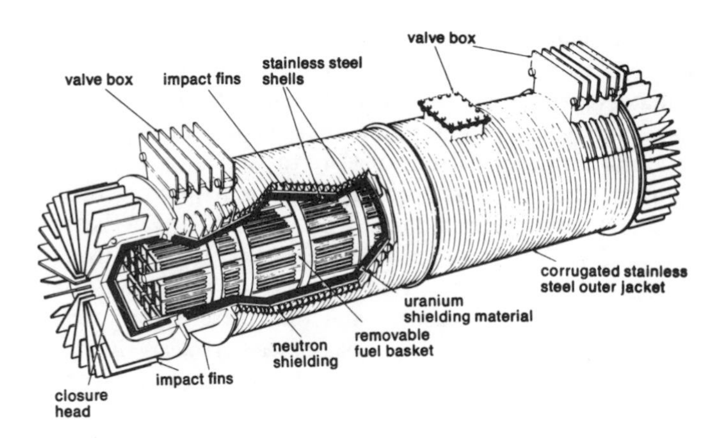

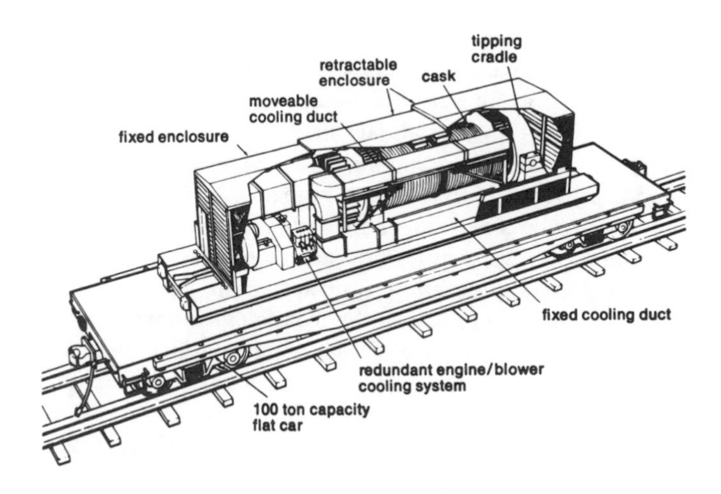

Target for the light-fingered. This is what thieves would have to steal (covertly!) to get at the fuel basket to get at the fuel rods to get at the plutonium oxide pellets inside. And that would only be the start of the long and difficult road to manufacturing a plutonium oxide bomb. (The figures above actually show the container and railcar to transport *spent* fuel; however, the general idea is the same for fresh fuel.) Vehicles that can be made impenetrable and immovable while automatically transmitting a radio alarm are now under development.

settle on the nuclear option in the first place, and then implement it in the power (rather than weapons) industry. The fact is that there are much easier ways of indiscriminately killing far more people. We shall mention some of them in connection with fossil-fuel storage, since this comes under the main framework of this book, but let it be said that even that is kid's stuff compared to other non-nuclear methods of wiping out people by the tens of thousands. What these methods are, a sufficiently determined terrorist will have no difficulty discovering, but he will have to do so without help from this book. We will merely quote Dr Cohen, who reports that "Experts on terrorism have stated that they hope terrorists will be attracted to nuclear plants as this might divert them from much more terrible things they could do more easily," and that these experts "consider the plutonium bomb publicity a great asset to society in diverting attention of would-be terrorists away from easier and much more harmful pursuits."[9]

THE MAD scientist who crafts a nuclear bomb in his basement, then, is stuff for Sunday supplements and Naderite disinformation; he is not a plausible threat. But what about the political terrorist with foreign support? What about the PLO obtaining a small tactical bomb, or at least instructions how to make one, from their Soviet buddies? What if they should threaten to set it off in Grand Central Station unless certain of their demands are met?

The prospect is improbable, but much less far fetched than Dr Frankenstein with a plutonium contraption in his garage. It is a prospect that cannot be prevented by technological means alone; it is a political problem that must be faced, not by politics, but by policy.

Not many years ago, the hijackings to Cuba took on epidemic proportions. Why have they stopped? The metal detectors, luggage inspections and air marshals have helped, of course, but they could never have done it by themselves. The hijackings stopped abruptly when Fidel Castro was persuaded to give assurances of extraditing future hijackers to the US. The learned dissertations by contemporary sociology professors to the contrary, the prospect of failure and punishment does deter crime.

The PLO and other organizations that kill indiscriminately have blackmailed their murderers out of prison in all concerned countries except one: Israel. Israel has, from the beginning, not merely proclaimed that it will not negotiate with blackmailers, but it has stood by its word, and terrorists have fared badly in kidnaping Israeli citizens or otherwise trying to blackmail the Israeli government.

On the contrary, West European governments have not lacked pompous rhetoric on the subject, but when the blackmail was on, they displayed all the spine of an overboiled noodle. Masquerading as "humanitarians," they caved in to blackmail in order to save a handful of lives today, knowing very well that they were encouraging the loss of hundreds of lives tomorrow and thousands on the day after. After hundreds of lives were lost to terrorists and blackmailers, Austrian Chancellor Kraisky still lives by the principle "Thou shalt procrastinate the killings until tomorrow," and there are only slight indications that the British, German and Dutch governments are beginning to realize that they cannot escape the political consequences of their ostrich policies.

The record of the US government has not been perfect in this regard, but it has been better than that of most democratic governments in that both the Nixon and Ford administrations have refused to negotiate with blackmailers over kidnapped diplomats. But it has done nothing to make known its determination — if, indeed, it exists — to take a hard line against nuclear blackmailers. Quite evidently the strongest deterrent against such blackmail is not the threat of punishment, but the threat that the effort will be ineffective.

From the much quoted book by Willrich and Taylor, we, too, will select a quotation:

"Moreover, such a 'hardline' policy may place the future development of nuclear power on more solid political ground. Such a policy would seem to insure that important democratic political values would not be gradually eroded in the process of adaptation to the ultimate security risks in nuclear power, if those risks materialize in fact. The government, in adopting such a policy, and the American people, in accepting it, would frankly recognize that no safeguards system will reduce the nuclear theft risk to zero; that the government and the American people are prepared to accept the risks involved in order to obtain the benefits of nuclear energy; but that they are not prepared to subject their political institutions to attack through nuclear bomb threats; ...*but* that they will insist on and support a safeguards system to keep the risk of nuclear violence as low as practicable."

Such a no-nonsense policy requires no technology. Its deterrent is not punishment, but failure: To quote Willrich and Taylor again, "Any person contemplating a threat would know that a major policy would have to be reversed for the threat to achieve his purpose."

The second powerful defense against terrorism — not necessarily nuclear — does not require any technology, either: infiltration of terrorist groups. Though not particularly singled out by Willrich and Taylor, it is a weapon that has proved its worth over and over again, and without destroying any civil rights. But it is a weapon that has, at the time of writing, been all but destroyed by an ideological group (usually mislabeled "liberal" or "progressive") in Congress and the press. They had no objections to the FBI infiltrating the Ku-Klux Klan, and neither had any other rational person. But when intelligence organizations try to protect us from the Fascists of the Left, this group wails to high heaven about civil liberties, as though civil liberties were of any use to the dead.

FBI Director Clarence M. Kelley recently stated that terrorism (though not necessarily nuclear) is on the rise and a major threat in the years to come. Yet the FBI and CIA are being bound and blindfolded.

The obvious conclusion is that the big boom to the terrorist is not plutonium; it is Senator Frank Church and his benighted admirers.

NOW FOR the jolly business of indiscriminately killing large numbers of people by abusing the facilities and fuels associated with non-nuclear power plants. Wherever there are high population densities and large amounts of stored energy in close proximity, there is an opportunity for terrorism, sabotage and blackmail.

The most obvious case is that of hydro-electric dams (they do not, of course, have to be hydro-electric). A dam failure killing 1,000 people has an average frequency of one in 80 years, higher (for the same number of fatalities) than major fires, explosions or chlorine releases;[11] in fact, except war, it is the most probable of all man-caused disasters with large loss of life.

The 1963 Vaiont dam disaster (p. 95) claimed 2,000 lives. Yet this was only an accident, not a deliberate act of sabotage with the express purpose of killing a maximum number of victims.

On the night from May 16 to 17, 1942, a British squadron of 19 Lancaster bombers attacked the Moehne, Eder and Sorpe dams in Germany and succeeded in breaching the Moehne dam, releasing some 30 million gallons of water into the Ruhr valley. There appear to have been several hundred dead, which was chickenfeed as air raid casualties went in those days, but that was not the purpose of the attack; the aim was to deprive the Ruhr (Germany's industrial heartland) of water and hydroelectric power.[12] The Sorpe dam received a direct hit, but the bomb hole was slightly higher than the water level. Albert Speer, the man who kept German production going through the air offensives, inspected the damage in the morning and writes in his memoirs: "Just a few inches lower — and a small brook would have been transformed into a raging river that would have swept away the stone and earthen dam. That night, the British came close to a success which would have been greater than anything they had hitherto achieved with a commitment of thousands of bombers... If the Ruhr reservoirs had been destroyed, the shortage of water for cooling the coke works and blast furnaces would have reduced production in the Ruhr district by 65%."

Yet this was an attack designed to paralyze industry, not to inflict heavy casualties on the population; moreover, to breach dams with pinpoint bombing by moonlight (radar was still quite crude) would have been a very difficult task even in the absence of the heavy German air defenses. But destruction of a dam above a populated area by high explosives in peace time involves comparatively little technical difficulty, and runs little risk of being foiled by the lax to non-existent security measures.

An act of sabotage against a dam could result in the death of 100,000 people; in fact, a 1974 University of California study on the risks associated with dam failures showed that there are several dams in the US where sudden failure could cause more than 200,000 deaths, and one of them could cause 230,000 fatalities. One of the authors of the report repeated these figures in a public hearing in 1976. (Once again, identification of these dams or the precise reference to the report would serve no useful purpose here, and they are omitted for obvious reasons.)

How many nuclear-physics PhD's are needed to make *this* nightmare come true?

IT HAS already been mentioned that an oil supertanker stores the energy of a two-megaton hydrogen bomb, but the really dangerous targets for terrorists would be the ships that bring liquid natural gas from Algeria and other places to US ports — sometimes only a few miles from the downtown districts of metropolitan areas (Boston, for example). Explosion of the cargo could release the total energy within a few seconds, and according to Dr Edward Teller, this could have the effect of a Hiroshima-type nuclear bomb.

It should be noted here that a home-made nuclear bomb would almost certainly *not* have the effects of a Hiroshima-type bomb. The two nuclear bombs in Japan were detonated high above the target cities, so as to deliver the blast and heat radiation to a maximum area. On the ground, a nuclear bomb would be a block-buster, not a city destroyer. And quite likely it would not even be that, for much of its effectiveness depends on the triggering mechanism. A nuclear bomb explodes by bringing two or more subcritical masses of fissile material together into a single bulk exceeding the critical mass. As soon as the first areas make contact, the subcritical parts must be held together *against the force of a nuclear explosion* for long enough to consume their unexploded fuel. If this is not done, the explosion will "fizzle," and the initial small explosion will simply blow the subcritical parts apart and back into safety. Thanks to the humanitarians who are so concerned about plutonium, we now have almost complete instructions on how to prevent that and how to assemble a nuclear trigger in our leisure time; however, the effectiveness of the explosion would be nowhere near that used in military hardware. The triggering mechanism of nuclear bombs is still one of the few remaining and closely guarded nuclear secrets, and it is not even known whether the Western powers, China, and the USSR use the same triggering mechanism.

The near catastrophies of Bayonne, N.J., in 1973, and in South Brooklyn, N.Y., in 1976, when thousands of people could have been asphyxiated in New York City except for the favorable weather situation (pp.88-91), can easily be precipitated by terrorists or saboteurs (though not by blackmailers). They merely have to wait until there is a temperature inversion and the wind blows in the "right" direction; there is not much to stop them setting an oil-storage complex ablaze covertly, and probably nothing to stop a group doing so by force.

This requires a little patience on the part of the terrorists; but nowhere near the patience needed (among other characteristics) to even try making a nuclear bomb.

THE REMARKS on terrorism and sabotage by means of fuels and facilities associated with non-nuclear power plants have been kept short because, as pointed out earlier, these considerations are largely irrelevant. Even so, it is evident that the dangers of abusing the non-nuclear power cycle for terrorism are greater than those of illicit nuclear bombs, for they are at least as great in their consequences, and far greater in their accessibility.

And yet the facilities and fuels involved in non-nuclear power generation are not the only, or even the most ominous, of the roads open to terrorists. There comes a point where analysis becomes a guide to action, and I will not pursue this ugly subject any further, since the point has surely been made. Let me just say that of the two conditions for indiscriminate mass killings, a high density of people and the release of large amounts of energy, only the first is truly essential, and that a high density of people does not necessarily require them to be in buildings. The erudite terrorist will know what I am alluding to, and it is my sincere hope that he drops dead before he attempts to try it out.

8

Reliability, Economy Conservation

Unsafe, unreliable, uneconomical and unnecessary.
Ralph Nader on nuclear power.

None of the subjects in the chapter heading above, quickly and exhaustively analyzed by Ralph Nader's expertise, are the subject of this book, since they are not intimately connected with the hazards of nuclear and non-nuclear power. Yet at least a few words ought to be said lest the impression arise that safety is the only superior quality of nuclear power, or that it is offset by other disadvantages.

Reliability is (outside the world of mathematics where it is a precisely defined term) a rather general concept, which is measured by certain indicators such as mean time to failure, availability (fraction of time for which a system is available), forced outage rate (duration of unscheduled down times to total time of observation), and several others. Large systems tend to be less reliable than small

Comparison of Fossil Units 390 MW and Above to All Nuclear Units

UNIT YEAR AVERAGES
1965—1974

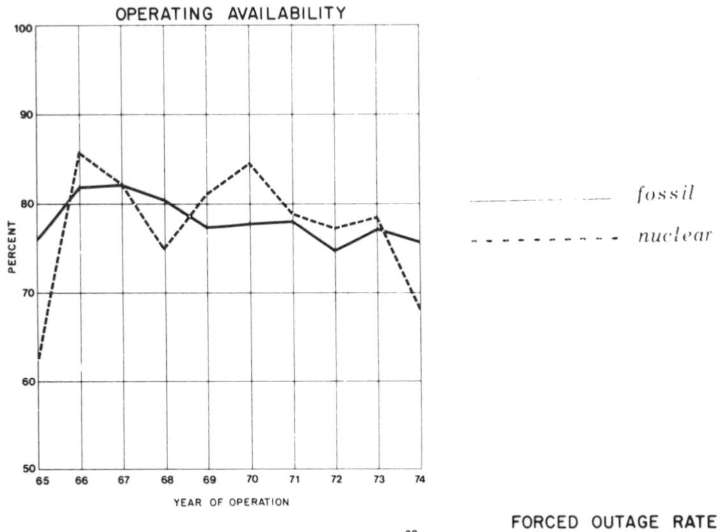

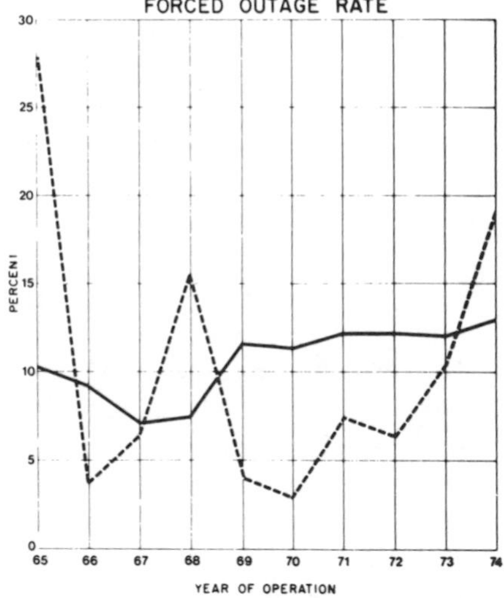

fossil

nuclear

From the Edison Electric Institute's *Report on Equipment Availability 1965 - 1974.* In the final effect, the reliabilities of nuclear and fossil-fired plants are about the same; but if fossil-fired plants were required to shut down for such puny deficiencies as nuclear plants, the reliability of nuclear plants would turn out much higher in the comparison.

ones; the reason is somewhat similar to the reason why a city has more cases of fractured legs than a family. New systems tend to be less reliable than moderately old ones; this is a phenomenon well known to every house or car owner — there is a "debugging period" before the system settles down to steady operation.

When comparing the reliability of nuclear plants to fossil-burning plants, one should therefore compare plants of roughly equal capacity. When this is done, nuclear power comes out with a reliability of the same order as fossil-burning plants, as shown by the figures taken from the latest available report of the Equipment Availability Task Force whose ten-year reports are sponsored by the Edison Electric Institute. There are years when nuclear plants do better than the others and years when they do worse; there are utilities whose nuclear plants have a far higher availability than its other plants (Commonwealth Edison of Illinois, Southern California Edison), and there are utilities with problem reactors plagued by particular deficiencies, sometimes even the subject of court litigations (Consumers Power Co of Michigan). But by and large, the two types of plant have about the same reliability.

Or so it would seem at first sight. In reality, of course, the reliability of nuclear plants is far higher, since they compete under incomparably tougher conditions. If all fossil-burning plants of the same type were shut down throughout the country merely because a hairline crack was found in the plumbing of stand-by equipment, the conventional plants could get nowhere near the reliability of nuclear plants.

THE diseconomy of nuclear power is a myth resting on yet another myth, namely that it is subsidized by the taxpayer. Yes, the American taxpayer has paid $1 billion to research nuclear safety, and I consider that a good investment; the American taxpayer also pays $1 billion, not total, but year after year, to Black Lung victims — not to cure or eliminate it, but just to compensate its victims. Nuclear power curbs Black Lung by striking at its cause.

Moreover, your friendly commercial entrepreneur, the US government, which runs such successful enterprises as Amtrak and the US Postal Service (the latter with an annual deficit of $1 billion), has very few enterprises that make money. But one of them is uranium enrichment, for which the fuel manufacturers pay through their noses, and another is Price-Anderson insurance, the premia for

KEEP IT COMING FOLKS, WE'VE GOT TO KEEP THESE PIGS HEALTHY!

From *Critical Mass*, March 1976. Though nuclear basic research was subsidized, commercial nuclear power receives no direct or indirect subsidies, but pays taxes and insurance premia; the government also makes a profit on uranium enrichment and on its part of the insurance fund. It is, on the contrary, foundations and organizations like Congressman Ottinger's *National Intervenors* (also using the name *National Coalition for Safe Energy*), whose income is not taxed and who receive tax-deductible contributions, although these funds are used "to help stop nuclear power" and therefore plainly "expended to influence the general public with respect to legislative matters" (IRS rule disqualifying deductible contributions).

National Intervenors
[illegible] Street, [illegible]
Washington, D.C. 200[illegible]

Yes, I want to help stop nuclear power. I favor safe energy sources such as the sun. Enclosed is my tax deductible contribution.

Name ....
Address ....
City, State ....
Zip ....

YES!

☐ $10 ☐ $25
☐ $50 ☐ $100
☐ $_____

Please make checks payable to National Intervernors/EAF.

which are paid by the utilities, partly to private insurance pools, partly to the US government. The private insurances pay first, and they have so far paid $400,000 for 26 minor claims; Uncle Sam hasn't paid anything yet (and probably never will), but sits on a fund of $8 million of as yet unused premia. And, of course, the utilities pay taxes — local, state and federal — with the stockholders paying a second round of taxes from their dividends. You call that a subsidy?

In any case, nuclear power is far more economical than fossil-fired power at present, and likely to remain so far into the future, for the price of uranium has little effect on the price of electricity (the biggest cost is the capital cost, the fuel cost is relatively small, as can be seen from the figure on the opposite page). At present, the cost of a kilowatt-hour of nuclear power is 50% cheaper than fossil-fired power in New England (where power plants depend heavily on imported oil), and 20% cheaper in the Midwest, where low-sulfur coal cuts down on the costs of pollution-control equipment. A little brochure called *How to calculate the costs of electricity* by I.A. Forbes[3] gives a simple way of finding the cost for all methods of power generation and any price fluctuations; the reader can check the economy in his own area and find the price levels to be expected in the future.

The ultimate test, of course, is whether the executives and the accountants of the utilities want nuclear plants, and they do want them very definitely. (Another incentive in the late sixties, now forgotten, was harassment by the environmentalists because of air pollution by fossil-fired plants.)

Isn't there something fishy about the whole issue when Ralph Nader is worried that the corporations won't make a big enough profit?

BUT aren't we running out of uranium ore?

We never run clean "out" of anything; the price of a commodity just rises as it is more difficult to obtain, until it no longer pays to produce or use it. In the case of uranium, lower and lower grade ore will have to be processed, which will drive up the price. But as a glance at the upper figure on the opposite page shows, nuclear power can take a doubling and even a tripling of the price of uranium without losing its competitive edge over fossils, even under the silly assumption that the price of fossils will not increase.

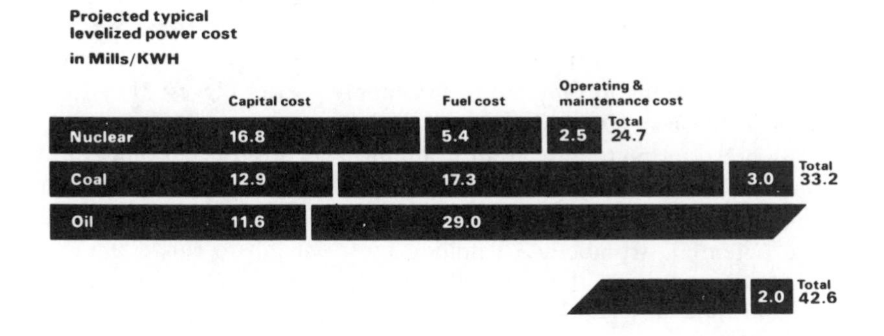

Costs of nuclear power compared to fossil fuels. 1 mill is 0.1 cent. (*AIF*)

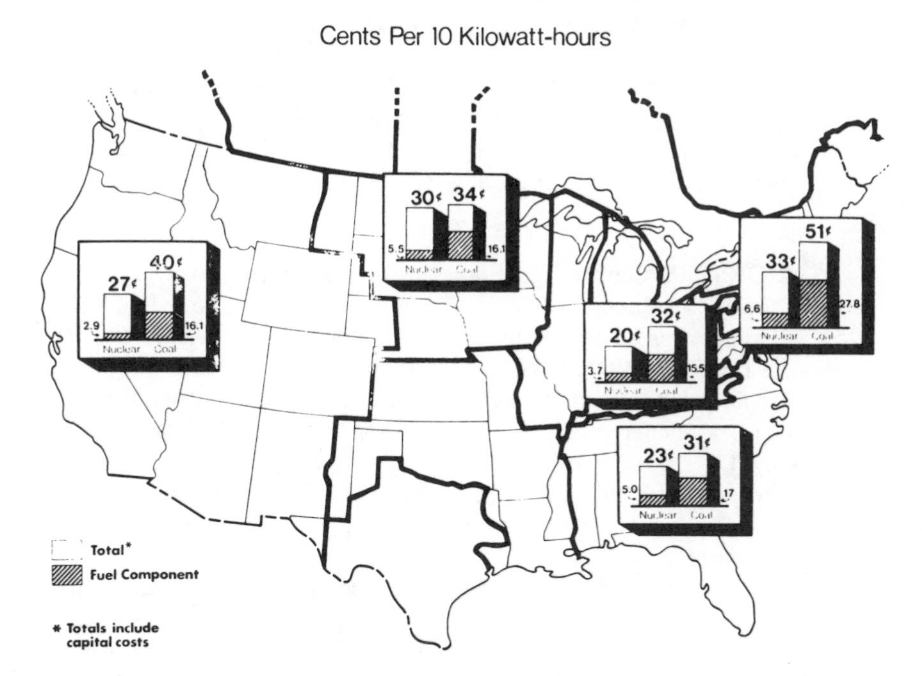

Costs of nuclear power versus coal-fired power by geographic regions (Edison Electric Institute, March 1976).

One should also beware of calculations based on "proved reserves." Proved reserves are the equivalent of inventory "on the shelf" in other industries. The proved oil reserves of the US, for example, amount to only a 11-year supply — the biggest they have been in the last 100 years.

By breeding plutonium from uranium, and breeding uranium 233 from thorium, fuel supplies for fission power can be extended to last not centuries, but *millenia*. The present uranium resources, proved and potential, are about 3.5 million tons, enough to run 800 Light Water Reactors for their full 40-year lifetimes; but if their U 238 content, now just going to waste as "tailings," were bred into plutonium, it could run those 800 reactors for *thirty-seven centuries.*[1]

However, only some 600,000 tons are *proved* reserves; as for the potential reserves, I will frankly admit that my knowledge of geology and other sciences needed to estimate them is lamentably close to zero, and on that point I am myself the one to ask "How is a layman to know?"

There are the three points made above (as well as some others), but for me the "clincher" is this: If it were true that we will run out of nuclear fuel by the year 2,000, would the profit-minded corporations and utilities want a technology that their geologists, financial and planning departments knew to be doomed to run out of fuel? I am perfectly willing to consider the charge that they are all crooks; but not that they are stupid crooks.

CONSERVATION is the answer to the energy crisis, say the anti-nuclear crusaders when sufficiently pressed as to what they would suggest as an alternative. Or solar power (which would use 50 square miles per 1,000 MW plant), or wind (which would supply a minuscule percentage if the entire country were plastered with windmills), or tidal energy (which would supply less than 1% if all usable sites on US shores were exploited).

In keeping with this philosophy, Nader has made it known that he bought a manual rather than an electric typewriter, and Comey boasts of not driving a car.

It would be easy enough to refute the argument that energy needs can be met by conservation alone; for in spite of all the waste, the US is still among the world leaders in efficiency of energy use when it is expressed as energy needed to create a dollar of GNP,[2] and there is just some fat to be trimmed before cutting into the muscle.

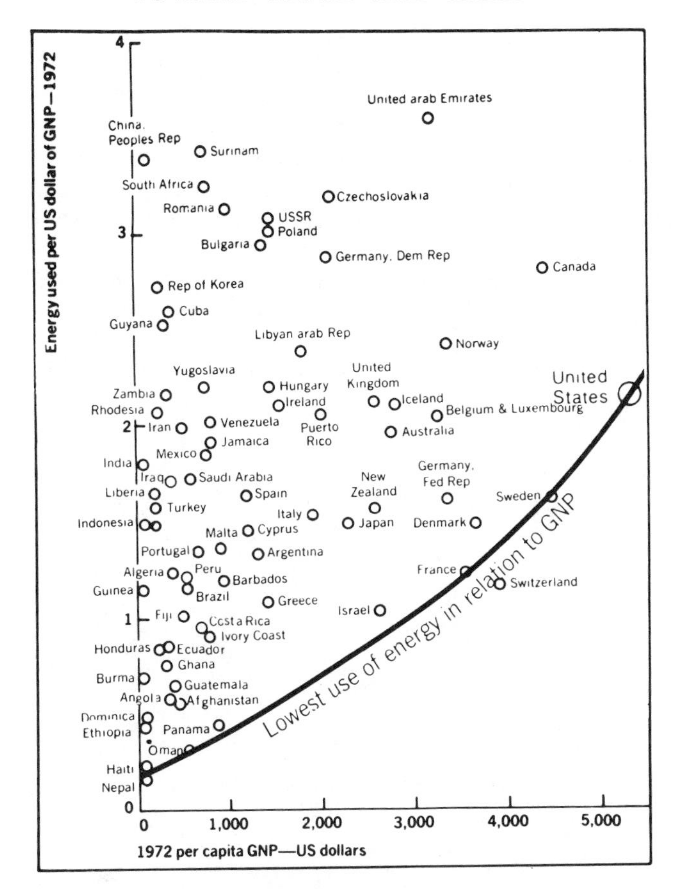

Efficiency in energy use. In spite of the indisputable waste of energy, the US still lies on the curve of lowest energy use in relation to GNP. [2]

But more to the point, the entire argument is irrelevant. Suppose indeed that Americans could be persuaded to give up their mobility and lifestyle (which they couldn't be), that the resulting mass unemployment would create no problems (which it would), and that energy consumption could be cut by half or more (which it couldn't); would it then be any less important to choose the safest possible form of electric power generation for the remainder of the consumed energy?

As we have seen, *per unit consumed energy,* the hazards to human lives are at least 100 times greater for electricity generated from coal than for nuclear energy; the risks of accidents are incomparably greater for fossils than for nuclear power, as are the risks in waste disposal; and the damage to the environment is about 10 times greater for coal than for orthodox fission, and about 1,000 times greater than for nuclear power using breeders.

Per unit consumed energy. That means that these ratios remain the same, no matter whether the total consumption is halved, doubled, or multiplied by any other factor. Conservation has not the slightest bearing on the point.

Yet Ralph Nader opposes nuclear power on the grounds of insufficient safety, while claiming that conservation makes nuclear power unnecessary. Consciously or not, what he is advocating amounts to thousands of needlessly lost lives, the spread of cancer and other diseases, and unnecessary rape of the environment.

In The Public Interest

By Ralph Nader

Attention William Simon, William F. Buckley, Milton Friedman, American Enterprise Institute, and other economic conservatives! Why are you so calm before the gathering storm of atomic socialism?

Where is your ideological fervor for free market enterprise when giant mismanaged corporations are pushing Uncle Sam (alias the small taxpayer) to bail them out by socializing their nuclear fission losses?

In case you want more details about this burgeoning corporate welfare system, consider the following:

1. The atomic power industry was launched in 1956 when the electric utilities demanded and received from Congress limited liabilit...

capital and uranium costs, and frequent shutdowns due to quality control failures or accidents.

It is burdened by serious and acknowledged safety problems that remain unresolved, such as where to put the deadly radioactive waste for the next 200,000 years. The entire $100 billion of the Rockefeller plan would be a drop in the atomic bucket. More like $1 trilli...

one being built at Barnwell, S.C. So, if they don't join the growing citizen movement against atomic power, consumers will continue to pay twice — as rate payers of their electric bills and as taxpayers for atomic socialism. This is assuming no atomic catastrophe that would result in untold casualties.

More evidence and more specialists are showing that atomic power is unsafe, uneconomic and unnecessary. Business Week magazine recently headlined a cover story entitled, "Nuclear Power Dims." Nuclear power also reduces overall job opportunities when compared with a program of energy conservation, and solar and other forms of practical, abundant energy sources.

9

Why?

Nuclear power is totally incompatible with human life and democracy. . . Reactor safety has been used, for the most part, as a red herring to preempt public debate.

Lorna Salzman, Mid-Atlantic Representative of the Friends of the Earth.

According to the facts and figures given in the preceding chapters, nuclear power, while not perfectly safe, is about 100 times safer than fossil-fired power. Why, then, should anybody object to it, and on grounds of safety to boot?

That is a far more complicated question than the simple figures indicating the health hazards of various forms of power generation.

First of all, is it possible that the ratio of 100 to 1 in favor of nuclear power is merely the result of distortions, omissions and perhaps even falsifications?

Hardly. In looking back over the past chapters, the reader will see that they are concerned with hard figures; not just with what *might* happen in nuclear power generation, but with the things that *have* happened and *are* happening in non-nuclear power production; with the results of reputable analysts and record-keepers, not political agitators. It is inconceivable that the record-keepers of lives lost in the coal cycle, or the scientific committees issuing radiological protection standards could be very far wrong; but suppose they were wrong by a factor as wild as 10 or 20 — then nuclear power would still be 10 or 5 times safer.

There are, of course, cases where scientists are more or less in the dark. For example, in air pollution, the relative amounts of hydrocarbons, nitrous oxides, sulfates, particulates and other pollutants are well known by weight; but it is a pretty nebulous matter as to what their relative impacts on human health are. Not so with the relative impacts of nuclear and fossil-fired power; the figures here are well known and not at all new. They have rarely been challenged by the nuclear critics, and when they have, the challenge turned out to be another display of scientific incompetence. But more often than challenged, they have been ignored.

These facts are, of course, well known to Ralph Nader and the other agitators; they have been pointed out to them often enough. Whether Nader and a score of similar agitators deliberately distort the facts or whether they are grossly incompetent is a question that may fascinate psychologists, but it has little bearing on the effect that they are having.

And, let's face it, that effect is considerable.

Why?

Because, some say, people love to be scared out of their pants; if they didn't, they wouldn't pay to see Frankenstein, Dracula or *The Prometheus Crisis*. Perhaps so, but this can explain only a small part of the phenomenon. People will pay to see a magician saw a girl in half only until they know he is sawing between one girl whose legs stick out of the box and another who is showing her head; once they know, it bores them. The nuclear box has been open for all who cared to look, so why would so many people prefer to swallow the superstitions instead?

Because, it is often said, the mass media artificially fan the anti-nuclear hysteria. That is certainly true and goes a little deeper, but not deep enough. The exaggerated coverage of the Naderite charlatans warning against nuclear power while the endorsements by qualified scientists and organizations are censored is a double standard so blatant that sometimes the media do not deny it; they defend their attitude by the old rule that "dog bites man" is not news, "man bites dog" is.

It is a false defense. If the media were only sensation seekers, but not otherwise biased, their coverage would be different. The Browns Ferry fire, in which no one was hurt and which never came close to any danger line, was "dog bites man" news, but it is harped on to this day. By contrast, the January 1976 oil fire in Brooklyn, quite apart from the genuine casualties, could have killed thousands of

New Yorkers if weather conditions had been unfavorable; had the media been merely sensation seekers, they would have presented it that way, but they didn't.

Was it mere sensation seeking when Edwin Newman (NBC) said that by the end of this decade America's rivers would be boiling, in large part due to nuclear plants? Does it happen every week that 33 outstanding scientists, 11 of them Nobel Prize winners, issue an appeal to the country? But the networks censored it, and CBS brought another of Nader's run-of-the-mill attacks on nuclear power on that day instead. Was it merely rejection of "dog bites man" news that made all three networks censor the news releases of the American Health Physics Society or of the American Nuclear Society? These organizations finally endorsed nuclear power, not after years, but after decades of painstaking studies, during which they adamantly refused to make such an endorsement; when they did finally endorse nuclear power, it was "man bites dog" news.

If the media were no more than sensation seekers, they could find plenty of people who claim that the earth is flat, or that we are about to be invaded by UFO-borne little green men, or that California is about to sink into the sea. They do occasionally report on such items, but they do not give it the repetitious and exaggerated coverage that they accord to the anti-nuclear charlatans, nor do they censor the opposing opinions. Why?

Because the media are not merely sensation seekers, but they are ideologically biased, and unlike the flat earth or UFO's, nuclear power has been made into a political issue.

A political issue? What does nuclear power have to do with politics?

Plenty. Not with small-time politics of Democrats versus Republicans, but with the bigger and deeper issues of ideological politics.

And here, at last, is one point we can agree on with the anti-nuclear crusaders, or at least those of them who have taken off the mask of environmentalism and have openly taken the issue of nuclear power for what it has become and what they have made it — a purely political issue.

Long before Nader's crusades openly took on the character of corporation-baiting binges and his annual Critical Mass rallies became reminiscent of the Nuernberg *Parteitags*, the so-called environmental movement developed heavily political and ideological overtones. I say "*so-called* environmental," because apart from the insane, nobody wants dirty air or foul water, so in that sense everybody is an environmentalist; but the leadership of the contemporary "environmental" organizations uses clean air and clear water only as a bait to mobilize the gullible for a far-flung campaign against "the corporate state," "big business," "vested interests," "the establishment," and whatever other devils they can wave before their faithful believers.

Even in the late sixties (let alone today) the ideological overtones and social origins of environmentalism were not hard to discern. Environmentalists *tended* to be against economic growth, for population control, against the Viet Nam war, for rapprochement with the Second (Communist) and Third Worlds, for greater permissiveness in legal and ethical issues, and for or against a wide range of other issues that had little or nothing in common with the physical environment. Those hardest hit by pollution, the poor, were conspicuous in the environmental movement by their absence. (How many chapters does the Sierra Club have in Harlem or Watts?) The typical environmentalist *tended* to be college-educated and affluent, and the movement was, and still is, strongest in the "information industry" — the media and the universities.

I have put the word *tended* in italics because I am not, of course, talking about one-to-one correspondences, but about statistical tendencies.

This group is sometimes lumped together under the name of "the liberals" — a misnomer, if ever there was one, for their attitude to liberty is ambiguous or hostile, and they are diametrically opposed to the liberalism of Adam Smith, John Stuart Mill or Friedrich von Hayek; while paying lip service to civil liberties, they strongly favor government interference and coercive legislation.

The urge for coercion and the arrogant premise that people do not know what is good for them are, in fact, two characteristics shared by this otherwise heterogeneous group.

These tendencies can, at times, become outright totalitarian. Paul Ehrlich, a population-controller, environmentalist and nuclear foe, probably considers himself a leftist radical and would object being likened to a fascist; but in writing on involuntary fertility control he states "Several coercive proposals deserve serious consideration, mainly because we [who is "we"? *P.B.*] may ultimately have to resort to them unless current trends in birth rates are rapidly reversed by other means."[1] That, surely, sounds more like an SS-Obersturmfuehrer than an American scientist.

"All scientists having a personal stake in the development of commercial nuclear power should disqualify themselves from the nuclear power discussion and leave the field to citizens who are perfectly capable of determining what endangers them and their freedom," writes Lorna Salzman,[2] exposing another totalitarian streak: She is quite willing to discuss the matter provided the opposition is silenced.

Other totalitarian trends have surfaced among this group of affluent malcontents, among whom environmentalism is just one facet. Violence "in the defense of higher values" has been tacitly condoned, and on occasions even actively supported. An extreme case (I hope) is that of Daniel Berrigan, an ex-priest, who testified in the defense of a murderer that "sometimes men must obey higher laws." (The man convicted of murder had planted a bomb in the computing center of the University of Wisconsin, killing a Ph.D. student, ostensibly to protest work done by the center for the Pentagon.)

The first unmistakeable signs of anti-nuclear violence "in the defense of higher values" have begun to appear: A pipe bomb was found in the reactor building of the Illinois Institute of Technology; dynamite was found at the Wisconsin-Michigan Power Company's Point Beach nuclear reactor; a break-in took place at the fuel storage building of Duke Power's Oconee facility in South Carolina; an incendiary device was detonated in a public area of the Boston Edison Pilgrim nuclear reactor; a fire, possibly arson, occurred in an equipment storage barn at Nuclear Fuel Services in West Valley, New York.

The Naderite and other organizations fanning the anti-nuclear hysteria, far from condemning these acts, publicize them (after extracting details from the NRC under the "Freedom of Information" procedure) as alleged illustration of how nuclear power is becoming a threat to public security.

In the spring of 1974, anti-nuclear activist Samuel H. Lovejoy toppled a tower at a planned site for a nuclear plant (it was a meteorological tower intended for measurement of weather conditions, i.e., exclusively for purposes of safety). In itself, this type of political vandalism is perhaps not very significant; what is significant is the way in which it is glorified by the anti-nuclear movement. A film called *Lovejoy's Nuclear War* was produced (there never seems to be lack of money for anti-nuclear propaganda), and there is talk of it being broadcast by the Public Television Network. Here is how one environmental journal[3] reviewed it: "What Lovejoy did was wrong, as is the reason why he *had to* [my italics, *P.B.*] do what he did." That is all too typical for the "condemnations" of violence by this group of alleged "liberals," whether it involves nuclear power or other issues that provoke their ire.

And these issues are many, most of them having the common denominator of free enterprise and the profit motive. It is by now trivial that environmentalism is being used as a crusading horse against the free-enterprise system. "This thin slice," says Ralph Nader, referring to the biosphere of the earth, "belongs to all of us" in an article called "The profits of pollution."[4]

Barry Commoner has a simple recipe for curing the energy shortage and pollution, both of which he alleges to be direct consequences of capitalism: Nationalize the railroads and all energy industries. "Economists and other students of capitalism," he writes,[5] "will recognize that the basic ideas I have discussed are those first put forward by Karl Marx... An explanation of why Marx's prediction [of the collapse of capitalism] failed to materialize — that is, until now — emerges from the improved understanding of economic processes which is a product of the recent concern with the environment."

"The real question we face," writes Physiology-Medicine Nobel Prize winner George Wald (whose many shameless distortions remind one that Nobel Prizes are not awarded for integrity), "is whether nuclear power can be produced safely while maximizing profits. The answer to that question is no."[10]

Lorna Salzman speaks of "nuclear power profit-making" and calls it "a technology that puts private profit and jobs over human health

The profits of pollution. (From *Critical Mass.*)

and lives, a technology which subdues man and his social institutions to technological tyranny."

And so forth — the point is too obvious to be belabored. Antinuclear attitudes tend (and I am still speaking statistically) to be voiced by people of a particular political persuasion. This has nothing to do with "guilt by association," in fact, it has nothing to do with guilt at all. It is simply a matter of statistical correlation.

It is a weird correlation, one that would have been thought utterly absurd ten years ago: If someone strongly opposes nuclear power, the chances are good that he admires Jane Fonda and dislikes John Wayne (neither of whom, to my knowledge, have ever voiced an opinion on nuclear power).

Like more illustrious writers, I am at a loss to give a name to this group of leftwing, college-educated malcontents who oppose nuclear power as part of a more general opposition to the "establishment." I roundly refuse to call them "liberals;" it would surely be an insult to John Stuart Mill to put him in the company of Lorna Salzman. Claire Booth Luce has called them the "America-stinks" crowd; Irving Kristol calls them "the New Class." In view of Barry Commoner preaching Marxism to the impoverished masses in the *New Yorker*, where they are also offered African safaris and $6,000 chess sets handcarved from walrus tusk ivory, perhaps a good name would be "the Penthouse Proletariat."

The Penthouse Proletariat is, of course, a very fuzzy concept. I cannot offer a definition, but here are two descriptions that may help to characterize it:

"DDT," writes Prof. E.N. Luttwak, "undoubtedly the greatest life-giving discovery of the century, is now a dirty word in exactly the same circles where the words CIA and Pentagon are dirty words... The same chorus tells us that we cannot and should not break the oil cartel, that we cannot build more nuclear plants because they are dangerous, that we cannot mine more coal because it ravages the earth, and lately, that we cannot drill for offshore oil because it would devastate the tidelands. At the same time, unemployment at home and the reduced ability of this country to feed the hungry are violently deplored, as if these were not the inescapable consequences of these core attitudes." [6]

"The ruling elites," writes Midge Decter in the same symposium, "...no longer have the conviction that the system, the civilization, is good and no longer wish to assume the responsibility of defending and cherishing it... I can't remember when I last heard [one of them] say a genuinely kind word about the system that made possible [their] own considerable elevation in it. But what I would say is that they are spoiled rotten and cosmically greedy. What ought to be the final object of their convictions, and a grateful conviction at that, they merely take for granted... Anything less than an uninterrupted flow of success, accompanied by an uniterrupted round of applause, they call evil. They have, blessed Americans, forgotten what evil is."[6]

This Penthouse Proletariat, or "America stinks" crowd, or New Class, or whatever you choose to call them, opposes energy use in general, and nuclear energy in particular. Powerful as the New Class may be ("They don't control the media," says Irving Kristol, "they *are* the media"), they are not strong enough to destroy the free enterprise system industry by industry and business by business. But they can and do strangle the lifeblood of them all, energy; they can and do go for the jugular.

They used to do tnis under the banner of environmentalism, for most energy pollutes and most energy is dangerous. Indeed, in the early days of environmentalism, when clean air and clear water still mattered, nuclear power was seen as the great hope of satisfying the environmentalists, for it does not pollute, its safety is higher than that of any other large-scale energy conversion, and it does negligible damage to the earth. Moreover, fossils are finite (they always stressed), hydroelectric power is running out of suitable sites (and is, like all energy facilities, doggedly opposed by this same crowd), oil, as the 1973 embargo showed, is vulnerable to foreign blackmail; but nuclear power comes from domestic sources, and with breeder technology, it can supply electric power for centuries (uranium) or millenia (thorium). Nuclear power, in short, did not fit into the campaign plan; its environmental impact was so small as to dismay the riders of the environmental crusading horse.

In part, the crusaders simply got off their horse; the environmental masquerade is dropped ever more frequently, and many antinuclear organizations, including Public Citizen, Common Cause, Critical Mass, Environmental Action, and the Friends of the Earth, are increasingly campaigning against nuclear power on purely political grounds.

That, of course, will not stand up to the facts, either; the alleged "garrison state" is a bigger exaggeration than thermal pollution ever was, and so the movement uses some more weapons borrowed from the totalitarian arsenal: the outright falsehood, and the lumping together of "enemies."

"Die juedisch-bolschewistischen Plutokraten" was the way Hitler lumped together his enemies in a lie. "The conspiracy of American imperialism and Chinese revisionism" is the Soviet phrase. "A technology that puts private profits over human health and lives," says Miss Salzman, lumping together the profit and safety issues.

I should perhaps interject here that I reject as absurd the "Communist conspiracy" theory; the idea of Ralph Nader being a Soviet agent not only strikes me as utterly absurd, but also as a gross overestimation of the sophistication and subtlety of the KGB thugs.* It is the Communists themselves to whose hearts the whole conspiracy concept is very dear; they see an "imperialist agent" under every bed and religiously believe that "the bourgeoisie" does nothing all day but study Karl Marx so that it might find the most fiendish tricks to delay its inevitable downfall.

Nor do I believe that the Penthouse Proletarians conspire among themselves, primarily because they do not need to. Not only do they achieve their destructive work by legal means, but they know very well what they are after and have little need of centralized leadership, let alone conspiracies.

And yet their totalitarian tendencies, remarked on earlier, have a characteristic that distinguishes left-wing from right-wing totalitarianism. There is obviously very little difference between fascism and Soviet communism; the two systems hate each other so much only because neither of them tolerates any opposition. But one difference does stand out, and that is the abuse of humanitarian ideals by left-wing totalitarianism. Fiendish as the Nazis were, they were amateurs in the art of deceipt; they made no pretense to oppose racism or to promote the Brotherhood of Nations, Social Justice or Human Rights. It fell to the Soviets to "improve" on fascism by concealing the same ends of unbridled power and brutal oppression in a mendacious cloak of humanitarian values.

It is this type (but by no means the same degree) of deception that is practiced by the nuclear crusaders, too. When Lorna Salzman

* Since the above was written, the Senate Internal Security Subcommittee made public a secret hearing[7] in which a former highranking Czechoslovak intelligence officer testified that the Soviet-directed spies at the Czechoslovak embassy in Washington (who make up more than half of the total staff) are under instructions to "heighten chaos" in America by any means, and that one of the spy agency's targets is Ralph Nader.

This has not changed my opinion expressed above. On the contrary, it confirms the crude stupidity of the Soviet establishment: Suppose that they were somehow to succeed in recruiting Nader as an agent (which I think highly improbable); what on earth would they expect him to do to "heighten chaos" that the man is not doing already?

charges that nuclear power is a technology putting profits over human health and lives, she cares not two hoots about human health or lives; if she did, she would support nuclear power instead of berating it. She must have seen the facts and figures of the threat to human health and lives by the alternatives, but she evidently dismissed them with the same mental acrobatics that enable the general member of the Penthouse Proletariat to ignore two million people driven at gunpoint into starvation in Laos, but to berate the South Vietnamese and South Korean governments for not living up to the standards of Jeffersonian democracy; to abhor South African *apartheid* while ignoring the murderous racism on the rest of the African continent; to condemn the torture chambers of Brazil, but not those of the USSR; and so forth.

"The decision to obtain 2% of our energy in barter for the human gene pool is morally indefensible and a national abdication of morality, writes Miss Salzman, this time masquerading in the cloak of morality. As of 1976, the nuclear share of total US capacity (the only pertinent standard) exceeds 8%, and unless Miss Salzman and her co-crusaders succeed in brainwashing the American people, that share is expected to reach 50% at the end of the century. As for genetic hazards, not even the nuclear bombs in Japan (let alone nuclear power) produced any observable genetic effects. The danger of genetic damage from nuclear power is so small that it cannot even be measured, but has to be estimated by indirect inference, and Prof. Cohen's estimate is "0.5 of a pant-hour,"[8] i.e., the damage to the gonads by routine emissions of a nuclear plant would have the same effect as wearing a pair of pants (thereby slightly raising the temperature of the testicles) for an extra *half-hour during a man's life*. That is what is morally indefensible to Miss Salzman; but her morality is strangely indifferent to the *Untermenschen* of Appalachia who are afflicted by Black Lung by the tens of thousands.

"The American public," continues Miss Salzman, "by broadening out the nuclear issue, will refuse to delegate its political power to a scientific elite..."

Note the future tense; she does not *hope* the American people will refuse, or *expect* it; no, she is the elected spokesman for the American public and with the confidence that the sun will rise tomorrow, she states that they *will* refuse. This *und-morgen-die-ganze-Welt* attitude is again characteristic of totalitarian propaganda, and its hidden purpose is the same. "The Soviet people will never diminish their vigilance against the imperialist spies and saboteurs" is a state-

—AP—24—

Signature Gathering Tactics

or, Sign My Petition Please

TABLE METHODOLOGY

Two people should work the table, one in front to bring the people to the table, and one behind the table, to make sure people sign properly.

Front person approaches customers – one on one. Make good eye contact and ask, "Are you registered to vote?" Customer will answer "yes" or "no". If yes, front person turns toward the table with a gesture and says, "Please sign to help get safe nuclear power." (Make it positive.)
Usually the person will move into the table at that point, so don't say anything else!

Person behind the table stands (alertly) with the pens clutched in his hand (not strewn around the table) and says, "Are you registered in ________________ county?"
The customer will say "yes" or "no."
If yes, hand him a pen, pulling him down toward the petition and say, "Sign the way you're registered to vote." THEN SHUT UP!

Most people will sign or not sign at this point – usually they sign.

Good Luck from People's Lobby

More deception practiced by the anti-nuclear crusaders: Here they need signatures, not facts. The California Initiative was purposely labeled "Safeguards Initiative," though its terms make it obvious that it is designed to shut down nuclear power.

ment that not only claims to make its author spokesman for the Soviet people, but one that seems to make itself vulnerable only as to whether the Soviet people will or will not diminish their vigilance; but in either case, it has smuggled in a lie, namely that the USSR is full of imperialist spies and saboteurs. Miss Salzman's cheerleading uses exactly the same technique: Her statement makes one doubt whether the American people will or will not do what she predicts, but in either case, she has smuggled in the lie that nuclear power necessitates a scientific elite to whom *political* power must be abdicated. This type of brainwashing draws the reader's attention from the pertinent point. (Have the American people delegated their political power by leaving all tooth extractions to an elite of dentists?)

"We should insist that they stop guaranteeing us perfect permanent operation of nuclear technology" is another of her statements that smuggles in a lie. Many readers, even critical readers, will ask themselves — should we insist? Shouldn't we? Should they stop guaranteeing? Shouldn't they? Which is alright with Lorna, for that will sidetrack them from the only pertinent question: When has a responsible scientist ever claimed that the operation of nuclear technology is perfect and permanent, let alone guaranteed it?

There is, however, one point where the anti-nuclear crusaders, as well as the Penthouse Proletariat in general, depart radically from the traits of hitherto known totalitarian ideologies, and that is their technophobia and science baiting. It seems that modern science and technology has wounded this new class of erudite intellectuals with the ultimate insult: They don't understand it any more. Every sentence of Lorna Salzman's little tantrum reeks of technophobia and hatred for science; what stares at the reader from between the lines is her gigantic inferiority complex, probably well deserved.

Now this is not only diametrically opposed to the traditions of fascism and communism, which elevated science to an esteemed position in order to abuse it for its own ends; it is also diametrically opposed to the professed concern of the elitists for the environment: For all but those capable of the weirdest mental acrobatics, it is axiomatic that to clean up the environment one needs more technology, not less. Automobile-caused pollution is not cleaned up by going back to the buggy and tons of horse manure, but by using lower compression ratios, ventilated crank cases, afterburners, catalytic converters, computer controlled fuel metering and ignition systems, and desulfurized gasoline; and later, perhaps, by switching from gasoline to methanol and other fuels cleanly synthesized from coal in processes that leave the coal in the ground without disrupting the surface of the earth. Pollution is a by-product of poor technology; not of technology as such.

Another front of the fight against the "establishment" is the ostensible Consumer Movement, which fights the system where consumers vote with their dollars in the market place and seeks to substitute a controlled, even regimented, economy instead. The Penthouse Proletariat is at least consistent in its muddleheadedness: The anti-nuclear crusaders oppose the safest form of energy on the grounds that it is unsafe; the environmentalists are against technology, the only hope of keeping the environment clean; and the Consumer Advocates are against the only system where the consumer is king (or would be, if a competitive economy were not made uncompetitive by the ICC, FEA, CAB, and hundreds of other tentacles of the federal bureaucracy). All three of these thrusts are personified in Ralph Nader himself, and whenever he wears the consumerist halo of this unholy trinity, he is, of course, advocating more tentacles, more agencies, more regulation and regimentation.

Dr A.L. Chickering[9] puts it as follows:

'The Consumer movement represents not actual consumers, but an abstract class of "Consumers" made up of upper middle-class reformer types (ostensibly) opposed to bourgeois values. Because "Consumers" don't like tailfins — whatever happened to tailfins? — they accuse businessmen of imposing "useless" (read "bourgeois") products on the public through advertising. The abstract class of "Consumers" oppose business not because business is indifferent to actual consumers, but, if anything, because business serves actual consumers *too well.* It's just that "Consumers" don't like the products that consumers want.'

Here again we have a trend that, if not totalitarian, is at least anti-democratic and anti-libertarian: It is not the genuine consumer who is to decide with his dollars whether a city needs a new department store, but the Consumer Advocates sitting on planning boards that legislate, regulate, regiment and re-distribute.

SO THE "new class" is a group of affluent, college-educated malcontents, highly influential by its dominance of the mass media and the universities. They masquerade as advocates of safety, as friends of the environment and as consumer advocates. They are opposed to nuclear power in particular, but to large-scale energy conversion in general. They exhibit totalitarian trends and favor legislative coercion over free choice by the consumer; they are hostile to the free-enterprise system, in particular to its backbone, the profit incentive.

All of which is a description, not an explanation. We have quoted Irving Kristol, Midge Decter, Edward Luttwak and Lawrence Chickering, who are among America's most brilliant writers on contemporary social and ideological affairs. But they, too, offer only descriptions, not explanations.

"We are left, then," writes Edward Luttwak,[6] "with self-destructiveness as an explanation. The dark and irrational motive forces of self-destruction are inherently refractory to analysis."

But that is only a description, too, and a capitulation to the difficulties of finding an explanation. Yes, history relates of many classes, groups and social strata that destroyed themselves. But none of them did so deliberately; their self-destruction was due to ineptness, miscalculation or complacency, but never due to deliberate intent.

There are, of course, individuals who go over the Niagara Falls in a barrel; but they do not build a fanatical following among a considerable section of the population.

And so we are still left with the question: Why?

There are, no doubt, tens or even hundreds of valid answers to that question, and every Penthouse Proletarian combines these motivations in a different mix of proportions. It would be naive to generalize or to present a single all-encompassing motive as the driving force between their seemingly irrational behavior.

Nevertheless, there are some motives more important and more widespread than others. Among the driving forces that I, for one, would *not* count as overwhelmingly important is the "vendetta" motive; it is certainly characteristic of the leaders of the anti-nuclear movement, but not necessarily of their followers. Ralph Nader is obsessed with the dream of making General Motors squirm; Kendall soured on the AEC when they did not, at first, sufficiently heed his warnings about deficiencies of the Emergency Core Cooling System, and he is still out to get their hide; Sternglass, Geesaman, Tamplin and Gofman all at one time worked for the AEC, and the utter rejection of their fantastic theories by their professional organizations only make them thirst more bitterly for revenge; the authors of *Power Over People* wrote their amateurish pap after a utility had obtained the right-of-way for a transmission line across their land; and so forth. Hundreds of others may be driven by the "vendetta" motive because they have a personal bone to pick. But there have always been individuals who were hurt, justly or unjustly, by some institution of the "establishment;" yet they did not succeed in raising a mass movement to avenge their grudge.

The thousands who rail against nuclear power, against the utilities, the corporations, the capitalist system and the profit motive have not been hurt personally by any of these; on the contrary, they have been the beneficiaries of the system they profess to despise. What can perhaps explain the actions of Nader or Tamplin cannot explain the motivation of the entire movement.

The hundreds of thousands in the radicalized sections of the American upper middle class may very well kid themselves that their opposition to nuclear power is based solely on safety and environmental considerations, and such high-minded rationalization makes it easy for this belief to become "sincere;" but objectively, this rationalization is pure nonsense, since nuclear power is demonstrably safer and environmentally sounder than any of its alternatives. Their convictions would soon be shattered if they were willing to look at the facts. But their convictions are para-religious, and they are willing to look at the "facts" only if supplied by the priests of the People's Lobby or the Friends of the Earth. There must be a strong motivation behind such irrationality, and it does not at all follow that the motivation is known to its victims.

But when it comes to motivation, look for the Number One motive of human action: self-interest.

What possible self-interest could there be in opposing the cleanest, safest and cheapest kind of electric power? What possible self-interest could there be in opposing economic growth in general?

Plenty. Perhaps the clearest indication is given by the population controllers, who as often as not are part of the radicalized, environmentalist, elitist, anti-capitalist, "America stinks" syndrome. Their beliefs are no less irrational than those of the nuclear foes. The Zero Population Growth movement emerged strongly in the late sixties, when the US birth rate had long been in an unprecedented rapid decline; their activity continued unabated when that decline took the fertility rate below the natural replacement level in 1973; and still it continues today when it is falling toward the point where not even immigration will make up for the eventual decline of the population (unless the 18-year old trend reverses). Concern about the population explosion in the Third World is but a red herring. Paul Ehrlich, the guru of the population controllers, may rave and rant about India, Bangladesh and South America; but when it comes to the punchline, he calls on the *United States* government to introduce incentives, if not coercive measures, to promote sterilization, abortion and birth control *in the United States.*

The credo of the population controllers is simple: "There are too many of you others." And the issue at stake is common with the seemingly wide range of other issues raised by the Penthouse Proletariat: They do not want to share their privileges with others. They are a class whose privileges are about to be lost if they have not been lost already.

What privileges?

The privilege of driving on roads with only a bearable number of other cars, driven by equally affluent and "cultured" people; a privilege already lost, for the roads are now crowded with the cars of the vulgar riff-raff that does not read the *New York Review of Books* and behaves as though they had an equal right to life, liberty and the pursuit of happiness.

The privilege of driving to the beach and finding it uncrowded. The privilege of driving rather than walking, and flying rather than driving. The privilege of flying without having to line up and rub elbows with the riff-raff. The privilege of getting away from the riff-raff by flying first class, for coach has already been invaded by it.

The privilege of being fawned over as the people who set the tone. The privilege of esteem accorded to those who drivel rather than produce. The privilege of living the good life of being sensitive, aware, concerned, involved and relevant, without being tainted by "materialistic" subjects such as physics, chemistry, engineering, business or dentistry, let alone by making the vulgar living of a plumber, electrician, printer, radar operator or (horror of horrors!) a wildcatter.

The privilege of being counted something better than "the others." It used to come with money. It doesn't any more, at least not without an awful lot of it. It used to come with education. It doesn't any more. The man who made $25,000 a year used to be somebody. So did the man who had written a dissertation on medieval Turkish literature. Not any more. Thrity-five percent of American youth, an unparallelled proportion anywhere or at any time, go to college. They worked just as hard as their parents (they mistakenly believe) to get a degree or to make it to the $25,000 level. But it does not give them the exclusive privileges that it bestowed on their parents. They can still fly to Florida in the winter, but they have to rub elbows with the riff-raff who do not go to analysts, who have not read the latest "in" novel on Lesbian incestuous rape, and who do not subscribe to the *New Yorker*, or even the *Saturday Review*. They have been cheated of their right to be somebodies; they have — almost — become nobodies like everybody else. For the nobodies now go to Florida, too. They even go to London, Paris and Rome.

No wonder the Penthouse Proletariat is frustrated. What caused this unnatural state of affairs? Who filled the jetliners with plumbers and printers? Who crowded the beaches with beer-drinking steel workers? Who made cars and gasoline so cheap that an entire nation was put on wheels? What gave people electricity at the laughable price of a nickel a kilowatt-hour? What let them make calls from California to New York for a silly dollar? What let a third of all American school children eventually pass through the gates of higher education?

Capitalism; science; technology. Stop it! Stop the world, I want it all to myself.

Here you have the self-interest of a class whose privileges are about to be cashiered. Not by legislation, not by oppression, but by the relentless advance of "the others," who want a better life, too. A privilege shared with everybody is no longer a *privilegium* or *priva lex* (private law); it becomes a *lex publica.*

I am not, of course, talking about the Rockefellers or the Kennedys; their kind of privileges are not yet threatened, and neither are they (except for some vote-garnering rhetoric by the family politicians) very ferocious environmentalists. I am talking of the upper-middle class intellectuals who lust after esteem and influence more ferociously than any robber baron ever lusted after money.

Nor am I claiming, to repeat this important point, that the self-interest of a group whose priveleges are threatened is the *only* explanation of environmental, and particularly anti-nuclear, irrationality. Obviously, the mix of deeper motives varies from case to case, and no single explanation can cover all the cases, nor even, perhaps, any single case.

But it does explain quite a few things. The fear of The Unknown explains some of the opposition, but not very much. It is true that people are less frightened of a dam break or an oil fire, even though both of these are more dangerous and more probable than a core melt, because they have a good idea of what floods and fires are like, but they usually have little knowledge of radioactivity. But how come, then, that the microwave oven is catching on fast, when people usually know less about microwaves than they know about radioactivity (and often think it is the same thing)? How come that they are swallowing tranquilizers, depressants, soporifics, pain relievers and hundreds of other drugs without accusing physicians of being a professional elite that puts profits over health, yet without an inkling of what exactly these chemicals do to their bodies?

The association of nuclear power with the first use of nuclear energy, the two bombs dropped on Japan, does not get us very far, either. Would people give up fire and the wheel because both were, at times, used as instruments of torture? Could Ralph Nader make them give up electricity by waging a psychological campaign that associates "electric" with "chair," just as he does with "nuclear" and "bomb"? Yet he has succeeded in turning thousands against nuclear power on the grounds that, paradoxically, constitute the very reasons for its superiority: safety, economy, availability, ease of waste disposal, and low environmental impact. What theory will explain the paradox?

The theory of the endangered class will. *Of course* the endangered class does not realize what their true motivation is. *Of course* they kid each other, and above all themselves, that they are motivated exclusively by legitimate humane concerns. Who didn't find wonderful and high-minded rationalizations to defend his privileges when they were threatened? To curb the power of ancient monarchies was to question the Divine Right of Kings and to go against God Himself, claimed those with a stake in the monarchy, and they probably believed it. To resist the New Order of the German Reich was to jeopardize the security of Europe, claimed the Nazis, and they probably believed it. To question the party line as decreed by the Politbureau is to be an enemy of the people, claim the Soviets, and probably they believe it. "There is only one politically, biologically and ethically acceptable solution," crows Lorna Salzman, "total and permanent abandonment of nuclear power." Conceivably, she believes it, too.

But once you discard what is claimed, assured and alleged, and instead watch the thrust of the action and its effects, the seeming self-destructiveness makes sense and the technophobia, the corporation baiting, the let-me-be-on-the-planning-board socialism, the fight against economic growth, the maligning of the profit motive, the professed hatred of "materialistic" values, the totalitarian trends, the systematic, destruction-bent harassment of private enterprise — they all fall into place. They are the actions of the somebodies who dread becoming nobodies.

Mass affluence is, by its very existence, destroying affluence as a distinctive sign of a favored social stratum. Economic growth, free enterprise and technology are the culprits who have committed this sin, and they must be stopped dead in their tracks. And they *can* be stopped by denying them their lifeblood, energy.

Energy generation can be drastically curbed by the legal roadblocks which this group so masterfully puts up via the allegedly environmental organizations, and people can be duped into cooperating by scaring them with horror fiction about safety, environment, corporate greed, even civil liberties (of all things!). And energy conversion does have drawbacks: It does have environmental impact, it does engender safety and health hazards, it does cost money, and much of it does have to be imported from unreliable sources.

Only nuclear power, while not perfect either, has less of these disadvantages than any other, and it has earned the hysterical hatred of the threatened class because it does not fit the campaign plan;

they wage war on it not *in spite* of its superior qualities, but *because* of its safety, its availability, and its economy.

"Nonsense," I hear the reader saying. "I know several anti-nuclear campaigners who are far from affluent and whose opposition to nuclear power is sincere and solely based on safety considerations."

And so do I. But apart from the fact that, I repeat, I speak of statistical tendencies, not exhaustively of all individuals (and apart from the fact that I also know people who are sincerely opposed to vaccinating children against diphteria), please look again. How sincere is a person (or what use is his sincerity) who has not compared the dangers of nuclear power to those of its alternatives? The fact that the campaigner may not be affluent means very little by itself; remember, Nader's raiders do not lust for money, they lust for power. Is he or she really interested in safety (as, no doubt, the campaigner asserts and believes), or are the arguments full of "big corporations," "participatory democracy" and the like?

And look deeper, for we know only a few people by personal acquaintance; look deeper for clues where most of the support and opposition comes from. Who is it that has so far most *effectively* opposed the anti-nuclear hysteria? Not the Atomic Industrial Forum with its sizable annual budget — you have probably not heard of them before. Not Westinghouse or Exxon — their advertisements were few and ineffective, perhaps even counterproductive. Not (I am sorry to say) the scientific community, which should have stood up to the brainwashing charlatans years ago. But whenever a chapter of the Americans for Energy Independence was formed, who was it that invariably gave support, not with money (that's easy), but with the actual work to be done? The International Brotherhood of Electrical Workers.

Let me say that I have, in general, little love for contemporary trade unionism, and I have a positive aversion against the American Federation of Teachers (which now recruits among university faculties and finds adherents mainly among those who could not make a living off campus and need protection of their incompetence). Yet I cannot help observing that when it comes to defense of nuclear power, the ones who speak up for it decisively are those who are nearest to the neutrons, the radioactive wastes and the emergency switches.

Who, on the contrary, is it that fans the anti-nuclear hysteria? The Sierra Club, whose members live in Beverly Hills rather than Watts, in Long Island Suburbia rather than Harlem; the Creative Initiative

THE BULLETIN OF THE ATOMIC SCIENTISTS
a magazine of science and public affairs

ences research laboratories. Most of the rest are in business, medicine, medical research, law, and government. More than 90% have B.A. degrees and 50% Ph.D.s or other advanced degrees.

Time Magazine has said that "The Bulletin's subscribers girdle the globe and can muster more scientific, diploma-tic, and statesmanship credentials than any world conference in Geneva." One study showed that 46% made trips out-side the U.S. within the two preceding years, and 75% travel by air annually. At the time of the study the average number of books owned by each subscriber was 438 volumes. Median family income fell in the top 5% range for all U.S. households.

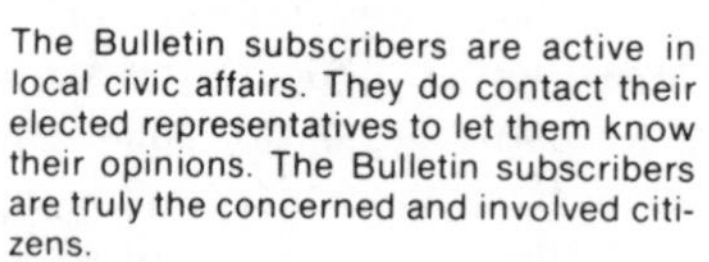

The Bulletin subscribers are active in local civic affairs. They do contact their elected representatives to let them know their opinions. The Bulletin subscribers are truly the concerned and involved citi-zens.

ADVERTISING RA

	1 time
1 Page	$500 00
2/3 Page	375 00
1/2 Page	300 00
1/3 Page	225 00
1/6 Page	125 00
Second cover	550 00
Third Cover	550 00
Fourth cover	600 00
Fourth cover, two color (publisher's choice)	650 00
Fourth cover, four color	700 00

Inside colors, inserts, bleeds, rates on r

CIRCULATIO

Total Distribution
Paid subscription
Newsstand and Other distribution

COMMISSION A
CASH DISCOUN

Agency Commission: 15%
Cash discount: 2% for paymen
of invoice date.

We, the people

Foundation, working out of its $500,000 headquarters in an affluent San Francisco suburb; Project Survival whose campaigners are the wives of high executives; the Radcliffe girls and Harvard boys who used to clamor for bussing until the day when their own children came to be bussed, and on that day they found themselves another ego-trip to go on.

And where was Lorna Salzman's little tantrum published? In the pages of Comey's Home Journal, the deceitful *Bulletin of the Atomic Scientists*, where the Sierra Club pays for full-page ads against nuclear power, where Nader and Kendall get the mailing lists for their gimmicky petitions, where the Rasmussen study is berated and *We Almost Lost Detroit* is extolled. Here Lorna says "The American public will refuse to delegate its political power to a scientific elite...," and she is far from being the only one to speak for all of the American public. But look deeper, not into the articles claiming that a nuclear power plant could melt a hole clean through the Earth or how America's energy needs can be met with windmills; do not look into the pages of this scare-peddling journal published by yet another tax-exempt foundation ("Educational Foundation for Nuclear Science") that speaks for the American people, but look at their advertising rate card (left), and you will see the American people staring right back at you.

Look deeper, and for all the exceptions (which I do not deny), you will see an affluent elite afraid that affluence of *everybody* could engulf them.

Some of the privileges of these upper-middle class elitists have been lost long ago and forever; the privilege of having colored servants, for example. Others are as good as gone — the empty roads, for example: The high-powered propaganda against the ostensible vulgarity of the automobile has not driven the "riff-raff" off the roads, though not for lack of trying. In yet other cases, they have shown themselves willing to sacrifice their privileges at the cost of denying them to everybody else, as in the proposals to forbid roads through national parks, which they would rather leave to the squirrels than to share them with the plumbers.

But one decisive privilege they still do have, and they will cling to it with the furious strength of desperation. They are still the tone-setters, public opinion moulders and educators. They still overwhelmingly control the mass media, the schools and universities, and they populate the obese federal bureaucracy. They still sit on all points of unelected power.

The "unelected" is not a fatal limitation, for uninhibited access to the ears and eyes of the voters can often program them for stands on ideological issues. Besides, there are cases when the citadels of unelected power can be stronger than the voting machines. Affirmative discrimination, for example, has been decreed by an unelected agency of the federal bureaucracy not simply ignoring the intent of Congress, but in diametric opposition to the anti-racist intent of the law.

Clearly, then, control of the brainwashing industry is no small asset to this elite whose time has come. They have been surprisingly successful in keeping the health hazards of the non-nuclear alternatives from the public; and they have been surprisingly successful in scaring the public not just with distortions, but with outright falsehoods.

It would be nice to finish this book by saying that in the end, reason will prevail. But will it? By the time you read these lines, the people of California will have voted on the "Nuclear Initative," whose contents is a cruel hoax. I expect this initiative to be passed. (A pessimist's surprises are always pleasant.) While other industrialized nations are forging ahead, while Britain, Germany and Japan are close to having breeders on line, and France already has had one on line for more than a year, orthodox fission power in the United States is being hounded and harassed to the point where its future is in doubt. The victims of the lack of nuclear power, the dead, the diseased and the crippled, pave the way for an ignorant "America stinks" elite lusting for power. The fact that reason has always prevailed in the past does not guarantee that it will prevail in the future.

Besides, it hasn't always prevailed. The glory of the arts and sciences in antiquity was buried for a thousand years by a doctrinaire and intolerant institution — the medieval Church — that considered science the work of the devil. For a thousand years, Western civilization was stifled in debilitating ignorance, poverty and backwardness.

It was held captive by an institution that had not come to power by the sword.

It had merely acquired for itself a monopoly on learning and the dissemination of information.

Notes

1. THE NUCLEAR MONOLOGUE

1. *San Francisco Chronicle,* February 12, 1976.
2. *San Francisco Chronicle,* February 5, 1976.
3. "Energy," *The New Yorker,* February 16, 1976.
4. P. and A. Ehrlich, *Population, Resources, Environment,* Freeman & Co. 1970
5. AIF news release, 10 Feb. 1976. Individual resolutions available from the organizations mentioned.
6. Declaration was presented to FEA administrator F. Zarb on 14 Nov. 1975.
7. Arlington House, 1976.
8. Follet Publ. Co., Chicago, 1973.
9. *Science,* 26 January 1973
10. *San Luis Obispo County Telegram-Tribune,* 12 September 1973
11. *Science,* 26 March 1976.
12. R. Wilson, paper given at Energy Conference, Center for Technology and Political Thought, Denver, Colo., June 1974. The figures will be slightly updated and modified in Chapter 3.
13. 200,000 in the universities, 200,000 in the federal government, 370,000 in industry. *1976 Statistical Abstract,* Bureau of the Census. (Data are for 1974.)

2. SOME BASICS

1. By J.G. Fuller, Readers Digest Press, 1975.
2. E.M. Page, "We did *not* almost lose Detroit;" free (in single copies) from Public Affairs, Detroit Edison Co., 2000 - 2nd Ave., Detroit, MI 48226.
3. B.L. Cohen, *Environmental Impacts of Nuclear Power,* Univ. of Pittsburgh, July 1975.
4. The radiation levels in this and the foregoing are based on *Nuclear Power and the Environment,* International AEC, Vienna, 1973; the US levels given there are, in turn, based on the 1971 estimates of the Environmental Protection Agency.
5. Lecture at the Los Angeles Chapter of the American Nuclear Society, 1974.
6. S. Jablon, "The Origins and Findings of the Atomic Bomb Casualty Commission," *Nuclear Safety,* vol.14, no.6, pp.651-659 (Nov.-Dec. 1973).
7. G. Farmer, *Unready Kilowatts,* Open Court Publ. Co., 1975, p.85.
8. All of these figures are taken from B.L. Cohen, *Environmental Impact of Nuclear Power,* Univ. of Pittsburgh, July 1975; Cohen's paper also gives the original references.
9. N.A. Frigerio, R.S. Stowe, *Report on very low level radiation,* Argonne National Laboratory, 1975; also N.A. Frigerio, K.F. Escherman, R.S. Stowe, *Carcinogenic hazard from low-level, low-rate radiation,* Argonne Nat. Lab., Report ANL/ES-26, 1973.

3. MAJOR ACCIDENTS

1. R.P. Hammond, "Nuclear Power Risks," *American Scientist*, vol.62, pp.155-160 (1974).
2. Letter to Project Staff Director of the Reactor Safety Study, 1 Nov. 1974 (published by Atomic Industrial Forum, 5 Feb. 1975).
3. Conference on Power Plant Siting, Portland, Ore., Aug. 25-28, 1974.
4. R. Wilson, W.J. Jones, *Energy, Ecology, and the Environment*, Academic Press, New York 1974.
5. L.B. Lave, L.C. Freeburg, "Health effects of electricity generation from coal, oil and nuclear fuel," *Nuclear Safety*, vol.14, no.5, pp.409-428 (1973).
6. Energy Conference, Center for Sc., Techn. & Pol. Thought, Denver, Colo., June 1974
7. See note 4.
8. "A study of base load alternatives for the NE Utility Systems," Arthur D. Little, Inc., July 5, 1973.
9. *Time*, 8 March 1976.
10. C. Starr and others, *Report to the State of California on safety of steam generating power stations*, Univ. of Calif. at Los Angeles, 1972.
11. I.A. Forbes and others, *The Nuclear Debate: A Call to Reason*, Energy Research Group, Boston, 1974.
12. Wilson and Jones (see footnote 4).
13. See note 12 to Chapter 1.
14. *Book of Facts*, Newspaper Enterpr. Ass., New York, 1976.
15. *Statistical Abstract*, Bureau of the Census, October 1975.
16. I am keeping this reference purposely vague. The intelligent reader will, after some searching of a better college library, be able to verify the statement and find the location of the pertinent dam. However, I do not wish to give easy directions to terrorists. (The latter are urged to try the manufacture of plutonium bombs; that will keep them busy and prevent them doing any real harm.)

4. WASTE DISPOSAL

1. *Environmental Impact of Nuclear Power*, Univ. of Pittsburgh, July 1975.
2. R.P. Hammond, "Nuclear Power Risks," *American Scientist*, vol.62, pp.155-160 (1974).
3. "Environmental hazards in radioactive waste disposal," *Physics Today*, pp.9-15, January 1976.
4. J.E. Martin and others, *Comparison of radioactivity from fossil fuel and nuclear power plants*, 91st US Congress, Joint Comm. on Atomic Energy, Hearings on effects of producing electric power, vol.1, pp.773-809, (1969).
5. Lave and Freeman, see note 5, Chapter 3.
6. D.J. Rose and others, *Nuclear power vis-a-vis its alternatives, chiefly coal*, M.I.T. 1975, as yet unpublished manuscript, privately communicated.

5. ROUTINE EMISSIONS

1. Countless times, e.g., Hearings before the Joint Committee on Atomic Energy, Jan.22-28, 1974.
2. "Nuclear Power Risks," *American Scientist*, March-April 1974.
3. International Atomic Energy Commission (see note 4, Chapter 2).
4. *Daily Camera*, Boulder, Colo.

5. L.B. Lave, L.C. Freeburg, "Health effects of electricity generation from coal, oil and nuclear fuel," *Nuclear Safety*, vol.14, no.5, pp. 409-428 (1973).
6. "Nitrosamines: Scientists on the Trail of Prime Suspects in Urban Cancer," *Science*, 23 January 1976.
7. Lave and Freeburg (see note 5).
8. National Academy of Sciences "Air Quality and Stationary Source Emission Control," prepared for the Senate Committee on Public Works, March 1975.
9. D.J. Rose, P.W. Walsh, L.L. Leskovjan, *Nuclear power vis-a-vis its alternatives, chiefly coal*, M.I.T., 10 Dec. 1975; preprint, privately communicated.
10. M.R. Rollins, R.W. Williams, W. Meyer, *Estimates of the economic effects a five-year national nuclear moratorium*, Univ. of Missouri, 29 Oct. 1975.
11. Wilson and Jones (see note 4, Chapter 3).
12. L. Lave, E. Seskin, *An analysis of the association between US mortality and air pollution*, Univ. of Pittsburgh Report, 1971.
13. J. Cairns, "The Cancer Problem," *Scientific American*, November 1975.

6. ENVIRONMENTAL IMPACT

1. R.P. Hammond, "Nuclear Power Risks," *American Scientist*, vol. 62, pp.155-160 (1974).
2. *Energy and the Environment*, Council of Environmental Quality, Aug. 1973.
3. Calculated from the data given in note 2.
4. *Nuclear Technology*, April 1975.
5. E.J. Mitchell, P.R. Chaffetz, *Toward Economy in Electric Power*, American Enterprise Institute, 1976 ($2 from AEI, 1150 17th St. N.W., Washington, DC 20036).
6. J.J. McKetta, *The world doesn't end here*, Council of Environmental Balance (obtainable free frm CEB, Box 7732, Louisville, KY 40207).
7. J. Maddox, *The Doomsday Syndrome*, Macmillan, London 1972.
8. The Fort St. Vrain plant in Colorado; the small prototype HTGR at Peach Bottom, Pa., finished its 10 year run in 1976.

7. TERRORISM AND SABOTAGE

1. Ralph Nader on PBS, WETA-TV, Washington, D.C., 25 Feb. 1975.
2. R. Lapp, *Nader's Nuclear Issues*, Facts Systems, 1975.
3. R. Lapp, *The Nuclear Controversy*, Fact Systems, 1975.
4. R. Lapp (see note 2).
5. G.W.Dolphin (member of Britain's National Radiological Protection Board), "Hot Particles," *Radiological Protection Bulletin*, July 1974.
6. *Comments on R. Nader's statements before the Joint Economic Committee of the US Congress, May 8, 1975*, Westinghouse Corp., July 1975.
7. B. Cohen, *The Hazards of Plutonium Dispersal*, University of Pittsburgh, July 1975.
8. *A 27 year study of selected Los Alamos plutonium workers*, Report LA-5148-MS, Los Alamos Scientific Labs., January 1973 (quoted by Cohen, see note 7).
9. B. Cohen (see note 7).
10. T.B. Taylor, "Proposed safeguard measures to assure against nuclear theft or sabotage," *Aware* Magazine, August 1974, quoted in Westinghouse Report (see note 6).

11. Rasmussen Report (see p. 71).
12. C. Webster, N. Frankland, *The strategic air offensive against Germany*, London, 1961; Sir Arthur Harris, *Bomber Offensive*, London, 1951.
13. A. Speer, *Inside the Third Reich*, Macmillan, 1970.
14. Testimony before California Legislature, Senate Committee on Public Utilities, Transit and Energy, 27 Jan. 1976.

8. RELIABILITY, ECONOMY, CONSERVATION

1. R. Lapp, "We may find ourselves short of uranium, too," *Fortune*, Oct. 1975.
2. F. Felix, "Where would we be without nuclear energy?" *Energy International*, October 1975.
3. Obtainable for $1.60 from Energy Research Group, Inc., 1661 Worcester Rd., Framingham, Mass. 01701

9. WHY?

1. P. and A. Ehrlich, *Population, Resources, Environment*, Freeman & Co. 1970.
2. *Bulletin of the Atomic Scientists*, November 1975.
3. *Organic Gardening and Farming*, December 1975.
4. *The Progressive*, April 1970.
5. *The New Yorker*, 16 February 1976, p.97.
6. "Has America lost its nerve?" (symposium), *Commentary*, July 1975.
7. *Communist Block Intelligence in the US*, Subcommittee on Internal Security of the Committee on the Judiciary, US Senate, 18 November 1975; Government Printing Office, Washington, DC (published February 1976).
8. B.L. Cohen, *Environmental Impacts of Nuclear Power*, University of Pittsburgh, July 1975.
9. "A Constituency for Business," *The Alternative*, Feb. 1976.
10. *New York Times*, 29 February 1976.

Index

This is the end of the book

but not of the nuclear controversy, nor of the energy crisis: the crisis that is not due to the lack of energy, but to the lack of access to it. For the access to it is blocked by government interference in free markets and by the roadblocks put up by the vociferous group described in the last chapter.

But this book dealt only with some aspects of nuclear power, and only with the situation up to early April 1976. There are new developments in all branches of the energy field almost every day. Will you rely on the Closed Fraternity (p. 26) to inform you fairly? Will you let yourself be browbeaten by the scientific jargon in which Barry Commoner clothes his advocacy of socialism?

If you have read this book carefully, you will know that Commoner's statement on the dangers of a nuclear plant, "There is only one other existing man-made device which has the physical capability of causing such a catastrophe in a single event — a nuclear bomb," is a flat falsehood, as is his statement that "nuclear reactors have a particularly low Carnot efficiency."

But will you be able to pinpoint his sophistries on coal, oil, gas, hydropower and solar energy? Will you know the facts that he conveniently left unmentioned?

Already *Business Week* is extolling his new book on energy, which advocates "reorganizing the US economic system along socialist lines" and is brimful of false technical assertions. And Professor Commoner is not the worst of the pseudo-scientific demagogues.

To protect yourself against the merciless onslaught of the brainwashing industry in matters of energy, subscribe to the pro-science, pro-technology, pro-free enterprise monthly newsletter *Access to Energy*.

It will keep you informed on the technical background of new developments in all branches of energy (not just nuclear), and that means not only primary sources, but also such facets as transportation, refining, exploration, new engines, transmission and storage, and a hundred others.

Access to Energy is written for laymen; yet PhD's keep renewing their subscriptions, too.

Access to Energy wrote about Ocean-Thermal Energy Plants long before other popular journals; it has reported on Migma Fusion, which has not yet reached the popular journals even now. It reports on flywheel vehicles; in-situ refining of coal; the potentials of methanol; it explains how solar energy can be genuinely harnessed, and how it can be used as a gimmick to bilk the taxpayer.

But its coverage is not limited to technical matters. The economic and political background of energy matters is reported, too. Who could have foreseen the Arab oil embargo in 1973? *Access to Energy* could, and did. Two months before the embargo, it warned of the folly of "importing ever-increasing amounts of oil from unstable sheikdoms in the Middle East, making the US ever more vulnerable to political blackmail."

Access to Energy is written and published by the author of this book, independently of the University of Colorado or any other institution. He owns no stock in any energy (or any other) company; *Access to Energy* accepts no advertising, no subsidies, no contributions; it is fiercely independent and beholden to nobody.

The only thing *Access to Energy* accepts is your subscription of $22 a year (early 1985 price) for 12 monthly issues. *YOU are the one* who will not renew if the newsletter is not good enough. You and the other subscribers, not the government, not the sham foundations, not the planning boards, but only the subscribers decide whether it should prosper or perish.

But it is not likely to perish. While other newsletters are folding right and left, *Access to Energy* is now in its twelfth year. The following are unembellished excerpts from some of the many readers' comments. They are spontaneous and *totally unsolicited*:

Access to Energy is a ray of sunshine amidst the gloomy news. I look forward to every issue.
P.A.H., Ann Arbor, Mich.

Most sensible, data-packed publication on the market. Excellently written, too.
D.O.H., St. Paul, Minn.

Your little journal packs more entertainment and educational value in one column than the other magazines put together. There is no equal in the field of energy and technology. Keep up the good work!
P.M., West Hill, Ont., Canada

My 13-year old grandson wanted to be a scientist, but was hesitant because of all the adverse propaganda he was exposed to at school — until reading you! That straightened him out, and he is very grateful to you, as well as being one of your most avid readers!
D.W.H., Costa Mesa, Calif.

And here are some of the letters congratulating *Access to Energy* on its 10th anniversary:

I receive a pile of mail every day. If I see AtE *in the pile, I reach for it first thing. It is both instructive and funny. How often do you find that combination? Only once a month, in my mail piles.*
Tom Bethell
Nationally syndicated columnist

I wish I had known about Access to Energy when you first published it. Two old quotations come to mind: "A voice crying in the wilderness..." and "You shall know the truth and the truth shall make you free." May you go on bringing us the truth as long as the Republic needs you.
Robert A. Heinlein
Author, Santa Cruz County, Calif.

With its stimulating and informative approach, your newsletter has contributed to public awareness of the importance of energy... You have my best wishes for many more years of Access to Energy. Ronald Reagan

I couldn't do without you. I have often wondered how the antinukes could survive the ministrations of your gentle criticism... Edward Teller

You take less risk than other new subscribers, for having read this book, you know the author's attitudes, convictions and style of writing.

If you think the newsletter is for you, please send $22 (or $1 in pre-1965 US silver coins) for a one-year subscription (12 monthly issues) to the address below. (Early 1985 price — subject to change.)

Thank you.

ABOUT THE AUTHOR

Petr Beckmann was born in Prague, Czechoslovakia, where he obtained his Ph.D and Dr.Sc. degrees. He worked for a research institute of the Czechoslovak Academy of Sciences until 1963, when he was invited to the University of Colorado and did not return behind the Iron Curtain. He is the author of 8 books and more than 60 scientific papers. Originally working in electromagnetics and probability theory, he became strongly interested in questions of energy and now publishes the monthly newsletter *Access to Energy* in his spare time.

Dr Beckmann is a Fellow of the Institute of Electrical and Electronic Engineers, a Registered Professional Engineer in the State of Colorado, and member of several professional organizations.

He has no personal stake in nuclear power, owns no stocks of any corporation, nuclear or otherwise, and is not involved in any research projects funded by any corporation or the federal government.